Indiana MacCreagh

The Real-life Adventures of the Original Indiana Jones

Roderick Heather

Published by New Generation Publishing in 2019

Copyright © Roderick Heather 2019

First Edition

The author asserts the moral right under the Copyright, Designs and Patents Act 1988 to be identified as the author of this work.

All Rights reserved. No part of this publication may be reproduced, stored in a retrieval system or transmitted, in any form or by any means without the prior consent of the author, nor be otherwise circulated in any form of binding or cover other than that which it is published and without a similar condition being imposed on the subsequent purchaser.

www.newgeneration-publishing.com

Other books by Roderick Heather

The Iron Tsar
Russia from Red to Black
An Accidental Relationship
The Summer of '56

"Man cannot discover new oceans unless he has the courage to lose sight of the shore."

"It is only in adventure that some people succeed in knowing themselves - in finding themselves."

Andre Gide (1869 – 1951)

Contents

Chapter One: Introduction 1
Chapter Two: Indiana MacCreagh 7
Chapter Three: An Alternative Back-Story 18
Chapter Four: Asia and Africa 35
Chapter Five: New York, New York 53
Chapter Six: White Waters 66
Chapter Seven: Three Men in a Boat 84
Chapter Eight: Ethiopia Once 104
Chapter Nine: Ethiopia Twice 132
Chapter Ten: Third Class Travel 141
Chapter Eleven: Project 19 163
Chapter Twelve: Home Again 175
Chapter Thirteen: Veni, Vidi, Scripsi 184
Appendix I: Gordon MacCreagh's Autobiographical Note 192
Appendix II: Alphabetical List of Gordon MacCreagh's works 194
Appendix III: Gordon MacCreagh's works by first date of publication 203

Maps

The Amazon Basin 70
East Africa and Ethiopia 108

Gordon MacCreagh in the Amazon, 1922

Chapter One

Introduction

Gordon 'Indiana' MacCreagh was an inveterate adventurer, intrepid explorer and big-game hunter, a self-taught ethnologist and entomologist. He was a keen photographer, an aeroplane pilot and a musician as well as being a competent sailor. He also had a working knowledge of several languages, including French, German, Spanish and Latin. By the age of twenty-three, he had lived in the USA, Britain, France, Germany and India, trekked across the Himalayas into Nepal, Tibet and China plus subsequently visiting Burma, Malaya, Thailand, Borneo and Africa. Later in his life, MacCreagh went on expeditions to South America and Ethiopia as well as travelling to many other countries such as Egypt, El Salvador, Canada, Mexico, Eritrea and Japan.

Apart from being a serial adventurer, MacCreagh was a prolific author and story-teller. In the period from 1913 to 1955, he had around 120 short stories of adventure and mystery published in various magazines and journals. In addition, he wrote various newspaper and magazine articles, almost thirty novelettes, one play and at least three non-fiction books, two of which became classics of their genre. Most of his short stories and novelettes were for publications in the USA but some were also printed in the UK and Canada. Literary masterpieces they were not, but his writing entertained and informed millions of avid readers for over forty years and it continues to be widely read and reviewed today.

MacCreagh lived and travelled in almost as many countries as the number of his stories and the itch to 'go-look-see' and explore what lay round the next corner was an integral part of his personality. He also had a deep sense of curiosity regarding the places he visited and the people that he met which served him well in his later life as an author. Many of the characters and plot-lines in his writings are clearly drawn from his own life experiences and as such make his tales of adventure that much more authentic and believable. MacCreagh essentially wrote as he lived; with an unbridled spirit and enthusiasm for life, mixed with a dry sense of humour and a willingness to tackle (and usually succeed at) almost any task that confronted him.

Although he had no formal academic qualifications or letters after his name, he was a well-educated and intelligent man. MacCreagh became a well-respected member of the Adventurers Club of New York, the American Society of Mammologists plus the Ends of the Earth Society. He was also awarded the Chevalier Star of Ethiopia by Emperor Haile Selassie. In later life, he deservedly achieved a national reputation as an ethnologist and anthropologist as well as something of a go-to expert on Ethiopian and Asian matters. He wrote many newspaper articles and lectured extensively across the USA on his life's experiences as an explorer and adventurer. Through his lecture tours, he also informed and educated the American public on significant world events of the time.

This book is an exploration of both MacCreagh's immensely interesting and exciting life as well as the possibility that he could have been the inspiration for the fictional Indiana Jones, as portrayed in the 1981 film *Raiders of the Lost Ark*.[1] For those who are not familiar with the film story, a short summary is in order. *Raiders of the Lost Ark* (also known as *Indiana Jones and the Raiders of the Lost Ark*), is an action, adventure film, directed by Steven Spielberg. The screenplay was written by Lawrence Kasdan, based on a story created by George Lucas and Philip Kaufman, with Lucas and Howard Kazanjian as executive producers with Stephen Spielberg as director. The film stars Harrison Ford in the title role and was the first instalment released of what became the Indiana Jones film franchise. *Raiders of the Lost Ark* is one

---

[1] Indiana Jones' full name in the films was Doctor Henry Walton Jones Jnr.

of the highest-grossing films ever made and was nominated for eight Academy Awards in 1982. The film was included in the National Film Registry of the U.S. Library of Congress, having been deemed culturally, historically, or aesthetically significant. The film's critical and popular success subsequently led to three additional films, *Indiana Jones and the Temple of Doom*, *Indiana Jones and the Last Crusade* and finally, *Indiana Jones and the Kingdom of the Crystal Skull*. A television series, *The Young Indiana Jones Chronicles* and fifteen video games were also produced later.

Set in 1936, the plot of *Raiders of the Lost Ark* essentially involves Indiana Jones, a university archaeology professor, being recruited by US Army Intelligence to search for and eventually recover the Ark of the Covenant[2] in Egypt. To do this, he goes through several perilous adventures in various countries and has to overcome his main adversaries, a thuggish group of Nazis who are also seeking the Ark. He is helped through his adventures by his enthusiastic girlfriend, Marion. The film persona is characterised by his appearance – usually wearing a fedora hat and carrying a bullwhip and leather satchel. He has a wry sense of humour, deep knowledge of many ancient civilizations and languages and a fear of snakes. His athleticism, inventiveness and marksmanship with a gun, somehow always seem to enable him to escape from any dangerous predicament that he encounters. In the back-story created for the later Young Indiana Jones series, the youngster travels around the globe as he accompanies his father on his worldwide lecture tour from 1908 to 1910. The story moves on to his activities during World War I as a seventeen-year-old soldier in the Belgian Army and then as an intelligence officer, seconded to French intelligence.

---

[2] The Ark of the Covenant is a legendary artefact made by the Israelites 3,000 years ago to house the stone tablets on which the Ten Commandments were written. It was described as being an ornate, gilded, large chest carried on poles inserted through rings on each side.

Whenever Lucas and Kasdan have been asked about the origins of the story, they have always made it clear that it wasn't based on any one particular individual. They say the film originated from Lucas' desire to create a modern version of the hugely successful pulp fiction adventure stories of the 1930s and 1940s.[3] The concept also followed in the footsteps of Hollywood's extensive catalogue of action and adventure B-movies that were so popular with film-goers in the 1940s and 1950s. The 1954 film entitled *Secret of the Incas*, featuring Charlton Heston, is a prime example of the precursors to Indiana Jones. However, it seems more than just coincidence that out of all the US states, Lucas and Kaplan should have randomly selected Indiana as the birthplace of their hero. One of them may well have come across the story of the real-life Indiana MacCreagh during their research for the film script and the mid-west location then stuck in their mind.

When comparing the fictional Indiana Jones to the real-life Gordon MacCreagh, the parallels in their respective stories are striking. The fact that, according to most sources, MacCreagh was born in Indiana is an obvious first link, as is the fact that he went on an expedition in search of the Ark of the Covenant (as it happens, in Ethiopia not Egypt as in the movie). In real life, MacCreagh was accompanied on this journey by his adventurous wife – not quite the fictional young Marion who helped Indiana Jones, but close enough. Although MacCreagh was neither a university professor nor ever employed directly by US Army Intelligence (at least, not that we know of), he did work for both the British and US military during World War II on a secret project in Africa. He also served as a pilot in the US armed forces in World War I. Although he never had to overcome German Nazis during his search for the Ark of the Covenant, he did come up against Mussolini's Italian fascist forces who were

---

[3] Called "pulp" because of the cheap paper they were printed on. Many of the stories were written in serialised format so readers had to buy the next issue to find out what happened next.

threatening to invade Ethiopia in the 1930s. And of course, his service in North Africa during World War II, when he was wounded, involved him in the Allied struggle against both the German Nazis and the Italians.

Although MacCreagh was not a professor, he became a respected ethnologist and explorer and like Indiana Jones, he certainly led a fascinating and extremely adventurous early life, travelling to some of the most remote places on Earth. He also visited Nepal and Peru, two of the countries in which the *Raiders of the Lost Ark* film is set. MacCreagh frequently experienced various life-threatening situations during his travels, whether it was being attacked by wild tigers and lions or by remote tribes with poisoned darts or spears. Over the years, he developed an affinity for and detailed knowledge of many of the primitive people that he encountered. Like Indiana Jones, MacCreagh usually wore long leather boots and a wide-brimmed fedora on his overseas expeditions plus he was a good shot, often carrying a German Luger pistol strapped in a holster on his right hip. In photographs, MacCreagh has the Luger in a cross-draw position, which indicates that he may have been left-handed. There are also some strong parallels with the back-story invented for the young Indian Jones – as we will see later. There is one small difference however, between the two characters – MacCreagh was fond of snakes whereas Indiana Jones suffered from ophidiophobia and was terrified by them.

Of course, the enigmatic MacCreagh isn't the only possible candidate for the real-life model on which Indiana Jones could have been based. At different times over the years, various other names have been put forward, including the American explorers, Roy Chapman Andrews, Hiram Bingham and William Montgomery McGovern, the mysterious German writer, Otto Rahn and even the British archaeologist and war hero, Lawrence of Arabia. But while each of them was a noted adventurer or explorer, none of them has quite the same strong links or parallels to the events in the Indiana Jones stories as Gordon MacCreagh. In addition, none of them claimed Indiana as their birthplace.

The fictional Indiana Jones was given a birth date of July 1, 1899, several years after MacCreagh was actually born, so the film's time-line and events do not entirely match those of MacCreagh's life. But that discrepancy applies to most of the alternative, possible real-life characters and overall, MacCreagh seems to have by far the closest fit. Also, the comparison is not just about facts, it's also about temperament and character. MacCreagh's constant desire to travel and experience adventure, to slip away from the real world, was a trait he shared with Indiana Jones. MacCreagh's often-sardonic humour and refusal to put up with those around him that he felt were fools, both place him close in personality to the fictional film character. Lastly, it only seems right to me that since MacCreagh wrote many of the pulp-fiction adventure stories which Lucas and Kasdan drew on for the Indiana Jones story, it is totally appropriate that he should be regarded as the original, real Indiana Jones. MacCreagh died well before the cinematic Indiana Jones was created and became so popular with audiences world-wide. Had he been fortunate to have seen the *Raiders of the Lost Ark* film, released in 1981, I think MacCreagh would have been intrigued and amused to encounter a character so like his own, facing events that partly mirrored the experiences in his own adventurous life. I would also like to think that this would have encouraged him to take up his pen one more time and give us his own real-life story. As the American writer and literary historian, Peter Ruber, said in his account of MacCreagh's life, "it would have topped the bestseller lists." [4]

---

[4] A biographical, on-line article entitled "The Incorrigible Explorer" published around 2002.

Chapter Two

Indiana MacCreagh

Those who are familiar with the back-story created for the cinematic Indiana Jones will know that his life and especially his childhood, are full of enigma and mystery; we never quite know all the details or the full truth. So, it is with the real-life Gordon MacCreagh. Little is known with any certainty about his early life and what facts are available often seem ambiguous or contradictory, often raising as many questions as they answer. If the following pages seem to read at times like a mystery story or a convoluted detective novel, rather than a normal biography, I readily apologise to the reader. Pulling together the different and conflicting pieces of information about MacCreagh's early life and trying to develop a believable and coherent account has not been an easy task. MacCreagh, intentionally or unintentionally, has left us with many conundrums and riddles.

According to most published biographical sources[5], MacCreagh was born to Scottish parents on August 8, 1886, in Perth, Indiana. Located a few miles north of the town of Brazil, Perth had been founded in 1870 as a speculative venture due to the proximity of the recently completed Indiana & St. Louis Rail Road which ran through the territory. The new settlement was named after its namesake in Scotland which was the birthplace of its founder's ancestors. Rich coal and iron ore mines had begun to be exploited in the area and these industries

---

[5] I have drawn primarily on three published accounts of MacCreagh's life – firstly, that written by Bob McKnight, one of his close friends, printed in the 1985 Chicago University Press reprint of *White Waters and Black*, secondly, the article by the American author Peter Ruber referred to earlier and lastly, the listing in Gale's Literary Resource Centre. Most other published accounts contain similar information.

attracted many people, including immigrants from Scotland. However, other than the tenuous Scottish connection, this small, mid-western township seems an unlikely place for MacCreagh's parents to have migrated to. MacCreagh's father, George, was a naturalist and historian who apparently came to the USA to pursue his interest in studying the culture and history of the American Indian.

Gordon MacCreagh's own writings provide no detailed information about his parents or wider family but we can assume that his father must have been a man with independent means. To have travelled to the USA to pursue his interest in the American Indian and to subsequently provide Gordon with an expensive education in Europe, would seem to indicate that the family must have been relatively well-off. We know nothing about Gordon's mother, Mrs MacCreagh, other than her Scottish provenance and there is no indication that Gordon had any siblings.

Once he was old enough, the young Gordon attended the local Perth public school for a few years. The simple brick and wooden frame schoolhouse, built in 1892, would have been almost new when MacCreagh started there. However, at the age of seven or eight, he was apparently sent to Britain to be raised and educated by his grandfather, a church deacon in Scotland. The reason for such a dramatic upheaval in Gordon's early life is not wholly clear. Some sources attribute the explanation as being due his father travelling extensively and often being away from home for long periods. Whether these absences were due to his study of the Native Americans or some other form of work is unknown.

Whatever the reason for being sent to Britain, young MacCreagh now experienced his first real adventure, a rail journey from Indiana to the USA's east coast and then a voyage by steamship across the Atlantic. On arrival in Scotland, Gordon presumably spent some time with his grandfather but apparently, soon found himself packed off to the prestigious Aldenham boarding school in southern England. Established in the late sixteenth century by Richard Platt, a master of the London Brewery Company, Aldenham School lies in the Hertfordshire countryside, some thirty miles north of London. It was a grammar school for fee-paying boarders and is not only one of the longest-established schools in Britain, but one of the oldest schools in the world. The choice of an expensive boarding school at the other end of the country is puzzling when there were plenty of good schools in Scotland.

After several years at Aldenham, MacCreagh was apparently then sent to Glenalmond School near Perth in Scotland which was founded as an independent school in 1847 by William Gladstone, later the Prime Minister of Britain. Originally known as Trinity College, it was established to provide teaching for young men destined for the ministry of the Scottish Episcopal Church and where they could be brought up in the faith of that Church. Its link to the Scottish Church was presumably the primary reason for the school's selection by MacCreagh's grandfather. Perhaps he hoped that the teenage MacCreagh might follow in his footsteps by joining the Church. There is no evidence to suggest that MacCreagh returned to the USA for any of his holidays during his time at boarding school, so presumably they were spent in Scotland with his grandfather. Being so far from home, away from his parents for so many years would have been difficult for MacCreagh but undoubtedly helps explain the strength of character, toughness and independent mind that he displayed in his subsequent life and career.

Aldenham School, England

Glenalmond College, Scotland

MacCreagh must have been an intelligent and able student at Glenalmond. Around the age of sixteen, he would have taken the higher level of the Scottish Leaving Certificate with exams in various subjects, including mathematics, a foreign language, Latin and science. He evidently did sufficiently well to allow him to progress on to higher education in Germany, probably at the University of Heidelberg.[6] Once again, the reason for this move to Germany is unknown. After some ten years of schooling in the British system, one would think that the natural progression would have been to a British university or possibly a return to the USA if either his father or mother were still alive there. MacCreagh's decision to pursue a university course in Germany seems to be a demonstration of his growing independence and the closing of a door in this particular period of his life. Just like the end of a chapter in one of the many stories that he would subsequently write.

Heidelberg University is one of the oldest and most prestigious universities in the world, having been founded in 1386. At the time of MacCreagh's arrival, the university was well known for its very liberal and open-minded spirit and widely recognized as a centre of progressive, democratic thinking. As such, it would have not only suited MacCreagh's developing attitudes to life but probably also helped shape his future thinking. For MacCreagh, exchanging the confines of his traditionalist, British boarding schools for the freedom and hedonism of university life in Heidelberg, must have been a welcome experience as well as something of an eye-opener. Some years earlier in 1880, Mark Twain had humorously detailed his impressions of Heidelberg's student life in *A Tramp Abroad*. He painted a picture of the university as a school for aristocrats, where students pursued a dandy's lifestyle, and described the great influence the student

---

[6] The US Who's Who 1934 entry for MacCreagh states that he was a student in Hanover and at university in Gottingen, not Heidelberg.

corporations exerted on the whole of Heidelberg's student life at that time.

Another famous author, Somerset Maugham, also studied at Heidelberg University, just a few years before MacCreagh arrived. His popular, semi-autobiographical novel, *Of Human Bondage*, published in 1915, has some intriguing parallels with MacCreagh's early life. The protagonist, Philip Carey, is orphaned in Paris at the age of six and is sent to live with his aunt and uncle who was a vicar in England. Carey's relations with his emotionally-cruel uncle rapidly deteriorate and he is soon sent away to boarding school. However, his uncle has a large library in which Carey discovers a love of books and the English language. He does well at school and has the opportunity to go to Oxford but rejects his uncle's preference and instead elects to study in Heidelberg. There is no direct evidence to suggest that MacCreagh either read this book or that the relationship with his grandfather followed a similar pattern to Somerset Maugham's auto-biographical character, Philip Carey. However, in one of his subsequent books[7], MacCreagh does provide this brief comment on his childhood, "all the stern teachings that had been beaten into me by my sainted grandfather." He was, apparently, a very devout Christian and could read the Bible in Gaelic. This is the only known reference by MacCreagh to his upbringing in any of his writings and tends to suggest that his life with his grandfather was not an entirely happy experience. Certainly, the course of events relating to MacCreagh's schooling, his subsequent choice of university in Germany and wanderings around the globe, do suggest a lack of closeness to any of his Scottish relatives.

According to most published accounts, sometime in late summer of his first year at Heidelberg, MacCreagh was provoked into a duel by another student. At the time, duelling with swords, known as academic fencing, was a popular

---

[7] White Waters and Black, published in 1926.

practice among German university students. It was seen as a mark of a young man's class and character, with duelling scars recognised as a badge of honour amongst upper-class Germans. Around the turn of the 20th century, Heidelberg University was a leader in this tradition and Mark Twain had even commented on it in his book, *A Tramp Abroad*. The approved method of academic duelling was for the combatants to stand facing each other and without moving or flinching, to accept whatever happened. Having the courage to take a possible blow in this way was felt to be more important than inflicting a wound on the opponent. In fact, the victor was often seen as the person who could walk away from the duel with an obvious scar. Students were expected to take turns with a thrust, primarily aiming for the unprotected parts of the face, especially the cheeks and forehead, rather than the body or arms. These duels were normally ended on the first significant loss of blood and to give point and kill an opponent was considered an assassination, with dire consequences.

The swords used were very sharp, rapier-like blades, not the light and flexible epées or foils typically found in fencing contests today. The duellists usually wore some form of protection on their necks, chests and plus mesh goggles for their eyes. Medical staff were often present but were not expected to do a good job of sewing up any wounds. As a result, thick, disfiguring facial scars were frequently the outcome of the contest and students were even known to pull open their fresh wounds and rub red wine into them to make them more noticeable. This whole process became so popular that facial scars became a trademark of German aristocracy in the early decades of the 20th century.

We don't know exactly what happened in MacCreagh's duel other than he apparently ended up driving his sword into his opponent's body rather than simply marking his face. Perhaps the etiquette or rules of the contest had not been fully explained to MacCreagh or maybe this was an accidental thrust due to inexperience. We do know however, that as the other student lay bleeding on the floor, MacCreagh thought that he had killed him. Fearing the consequences, both within the university and more seriously, a potential action in the German courts, MacCreagh quickly fled the scene and laid low for a while. He had only just turned eighteen and must have been in a panic, terrified and at a loss as what to do or where to go. There was no time to contact his parents or grandfather to ask for help or advice about the serious situation in which he now found himself. Probably for the first time in his life, he now had to take full responsibility for his future and it is significant that he chose to neither return to Scotland nor to the USA. Instead, he packed his possessions and leaving Heidelberg in the middle of the night, he eventually headed south to begin the next chapter in his life in far-away India. MacCreagh later claimed that as a young boy, he had always wanted to see the world. He had already seen something of the USA and Europe. Now was his chance to explore further afield. The termination of his period of formal education in such a dramatic and unexpected manner proved to be the catalyst that would propel him on to new adventures and provide the foundation for his future career. Although MacCreagh was to learn much later that his duelling adversary had recovered from his wound, the die was now cast.

Although the above version of MacCreagh's early life is intriguing and dramatic, it doesn't stand up to detailed scrutiny and leaves several key questions unanswered. Firstly, why did his parents decide to emigrate from Scotland and settle in Perth, Indiana? This relatively remote, mid-West location in Indiana's Wabash valley doesn't seem a particularly convenient or natural choice for the kind of activities that George MacCreagh was supposed to be intent on pursuing. For a man with a scholarly or intellectual bent, settling down in this small, unsophisticated community of miners and steel workers was an odd choice. With only a few hundred inhabitants, no rail-head or post office, no cultural activities to speak of, no library for reading or research and distant from any areas of significant interest for studying Native Americans, Perth seems to have had nothing to offer the MacCreagh's. It is possible that they had close friends or relatives among the recent Scottish immigrants already living in Perth. However, if they did, one can see Perth as a starting point for the MacCreagh's as newly arrived immigrants but not as a place for them to put down roots and raise a family.

Secondly, what really happened that caused the young MacCreagh to be sent to Scotland? The decision to send Gordon away to be raised by his grandfather in Scotland could not have been an easy one to make and suggests other factors at play. In normal circumstances, one might have expected Gordon to be looked after by his mother at home in Perth, even if his father was regularly away. The situation could be explained by the break-up of the MacCreagh's marriage or possibly, the untimely death of one or both of his parents. There are two facts that lend some credence to this hypothesis. Firstly, Gordon was an only child at a time when most families had multiple children which suggests a marriage cut short in some way. Secondly, it appears he never returned to the USA to see his parents during his ten years of schooling overseas, nor indeed during the ensuing ten years of his early working life. It seems inconceivable that, if one or both of his parents were still alive in America, he would not have made some effort to return and visit them.

Whatever the reason for being sent to Britain, it seems highly unlikely that Gordon would have made such a momentous and complicated journey on his own at such a young age. Someone must have accompanied him on the long trip from Perth to Scotland. If the MacCreagh's marriage had failed or if either parent had indeed died prematurely, then it is possible that Gordon travelled to Britain along with his surviving parent. However, since there are no records of any MacCreagh departing the USA or arriving in Britain at this time, the likelihood of this transatlantic journey having actually occurred is open to doubt.

Thirdly, who made the unusual choices of schools and university and where did the money come from to pay for the significant costs of his time at exclusive boarding schools and university? There is no hint of an explanation in any of MacCreagh's writings as to the source of this funding. His grandfather's position in the Scottish Church was a lowly one and could not have provided the necessary income. Also, whatever skills his father, George, had as a naturalist and historian, it seems highly unlikely that such a profession could have generated the money required to pay for the high-level education provided for his son. If George was sufficiently wealthy to have emigrated to the USA and then funded his son's private education over many years, one would expect to find some footprint in history of him, either in Scotland or the USA. Yet there is none. There is no record of any George MacCreagh being active in the fields of nature or history, nor indeed in any other intellectual or commercial activity during this period. There is no evidence in immigration records of a Mr or Mrs George MacCreagh entering the USA in the latter half of the 19th century and there is no mention of either of them in any US census records. Most importantly, neither of the British schools mentioned earlier has any record of a student with the name of Gordon MacCreagh (nor any similar surname) and neither does Heidelberg University. So why do the accepted versions of his early life story include these specific details about his education and where did the information originally come from?

The answers to these vital questions are that some or most of the above version of MacCreagh's early life is pure fabrication and invention, presumably mostly originating from the author himself. We have to dig deeper and do a little detective work to try to find an alternative, more credible version of our Indiana MacCreagh's early life. If we want to understand the man and his character, we also need to appreciate how and why these riddles in his early life story arose.

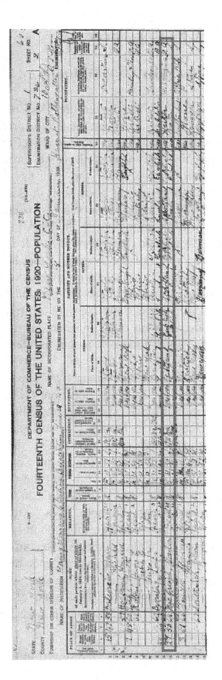

US 1920 Census Form

Chapter Three

An Alternative Back-Story

The first known factual record concerning MacCreagh's life dates from early 1920. In that year, the US government conducted a national census and Gordon MacCreagh, reported as a New York resident at the time, provided some revealing information. On the census form, MacCreagh is recorded as having been born in Scotland (estimated as circa 1890), along with both his parents. It also shows that he was an immigrant to the USA in 1911 and then became a naturalised American citizen. It is hard to think that this was a slip of the tongue, or that he was confused by the questions. There seems no coherent reason for MacCreagh to have provided false information about his origins to the census officer and the date of arrival in the USA of 1911 ties in with other information that we will consider later. This US census record is clear and unequivocal.

However, just one year later, in January 1921, we have a very different account. In his application for a US passport in preparation for his upcoming expedition to South America, MacCreagh curiously gave his place of birth as Perth, Indiana and most perversely, that his father's place of birth was Chicago, Illinois not Scotland. He also entered 1889 as his year of birth, not the date of 1886 that features in most accounts of his life, including his own, brief autobiographical note.[8] An extremely curious change of 'facts' in just twelve short months.

Since these are both official documents, the conflicting information they contain has to be given credence but it is hard to assess which parts are correct. There are several other later records that support the 1889 date of birth, including passenger lists of ships on which MacCreagh

---

[8] Published in the *Argosy* magazine in 1933 as part of a series about their authors.

subsequently travelled and business directories or surveys for Manhattan in the 1920s. So, I feel the date of 1889 is in fact the correct one and, for the remainder of this book, I have used the year of 1889 for MacCreagh's birth, not 1886, when referring to actual dates and events. However, there is sadly, no other solid evidence to confirm MacCreagh's actual place of birth. The name of Gordon MacCreagh doesn't appear in any of the available birth records for either Indiana or Scotland.[9] The same is true for his father George, with no mention of his birth or death in either Scottish or American records. Given all the information to the contrary in published sources, MacCreagh's statement that his father was born in Chicago seems to be impossible to explain or understand.

If Gordon MacCreagh had in fact been born in Perth, Indiana, he would normally have needed a US passport when he was sent away to Britain for schooling. Since MacCreagh's 1921 passport application appears to be for a completely new document, not a renewal, it indicates that he was never issued with a US one as a youngster. Whether he had a passport or not when he arrived in Britain, MacCreagh most certainly would have required some form of identity papers for his subsequent journeys to Germany, Asia and Africa. If he didn't possess a US passport, he must have travelled using a British one. This tends to confirm the US census information that MacCreagh was not born in the USA but somewhere in Scotland. Unfortunately, we don't know what identity papers he had for his subsequent overseas travels but as a British citizen, it would certainly have also been much easier for MacCreagh to subsequently take up employment in India.

---

[9] Indiana State did not require formal registration of all births until the early 20th century. However, it seems unlikely that MacCreagh's father, an immigrant and educated man, would not have registered the birth of his first child.

MacCreagh's Passport Photo

Owen Cattell (on right)

MacCreagh's 1921 Passport Application

There is one significant, additional piece of interesting information in MacCreagh's passport application. He was required to produce a birth certificate with his application and on the form, there is an official stamp with three words which appear to state "birth certificate seen". This suggests that MacCreagh must have submitted some kind of certificate to the officials. Interestingly however, in a formal letter accompanying his application, he apologises for "the fragility of the certificate of birth" provided and that it is not fully legible as it is stuck to another sheet of paper "on which it was forwarded to me." Where did this information come from, who gave it to MacCreagh and when? Given his use of the word "forwarded" it implies it must have been sent rather than given in person, possibly either by one of his parents or his grandfather in Scotland? Evidently, it was sufficient to confirm to the government official that the information in MacCreagh's application about being born in Perth, Indiana on August 8th, 1889 was correct as well as presumably, his father's birthplace of Chicago. It all sounds rather suspicious and one is left wondering about the source of the certificate produced and whether it was actually genuine.

In this connection, it is worth noting that MacCreagh's passport application was witnessed by a man called Owen Cattell, a photographer, apparently resident at the same New York apartment block as MacCreagh. Cattell was an interesting character in his own right and seems a strange person for MacCreagh to have chosen as a witness. Cattell was the son of a wealthy but egocentric professor of science at Columbia University whose own career was full of controversy. In 1917, as a seventeen-year old student at his father's university, Owen was arrested and convicted, along with a couple of other students, of conspiracy to obstruct U.S. conscription laws during World War I. The case was widely reported in the New York press. It therefore seems odd that MacCreagh chose the notorious Cattell as a witness, especially as MacCreagh had volunteered for the US armed forces in World War I.

Later, Owen followed his father into the world of science and like father, like son, he also proved to be an extremely unreliable and difficult man to get along with. In his witness affidavit for MacCreagh's passport, Cattell said he had known him for more than five years which infers that he met MacCreagh when he was only fifteen or sixteen and still at school. This seems highly implausible and when taken together with the dubious character of Cattell, casts doubt on the veracity of the information provided in MacCreagh's passport application as well as the authenticity of his birth certificate. Since MacCreagh and Cattell between them combined the skills of photographers and artists, it is possible the certificate was an inventive forgery.

MacCreagh's passport photograph is the earliest known picture we have of the man. Although passport photos are notoriously unflattering, it shows a very earnest-looking Gordon with a prominent forehead, straight nose, slightly wavy brown hair and piercing, grey eyes. Behind the serious look however, is just the suggestion of a smile, hinting at the author's indomitable sense of humour. He was described as being of medium height at 5' 7" with various scars on his forehead and chin – presumably obtained during his many encounters with wild animals while in Asia and Africa.

Although MacCreagh did not attend the two boarding schools in Britain as described earlier, there is very strong evidence to suggest that he did in fact spend time in a British school somewhere. He clearly had knowledge of the British school curriculum[10] and many of his stories, especially the earlier ones exhibit a familiarity with English idioms plus a language style and vocabulary that is at times, more English than American. He also knew Latin which was a subject much more likely to be taught in a British school than in Perth, Indiana. The fact that he was

---

[10] For example, see page 10 in the 1928 edition of *The Last of Free Africa*.

already an able player of Scottish bagpipes when he first arrived in New York also seems to suggest a familiarity with Scottish life during his formative years. Wherever MacCreagh went to school, he evidently had a decent education and was an avid reader. His grasp of the English language and history plus the references to Greek classical legends and French authors in his writings attest to that. Although it seems he wasn't a pupil at the Glenalmond school, it is located in Perth, Scotland and if that is where MacCreagh was really born and raised, he would probably have known of it and thus selected it as 'his school'.

If we now jump forward a few years, there are a couple of press articles that shed more light on MacCreagh's early life - the April 1st 1927 issue of North Carolina's *High Point Enterprise* newspaper and a vividly illustrated article in the Illinois *Mattoon Gazette* printed in January 1933.[11] They both appear to be based on contemporaneous interviews with MacCreagh and both versions are similar, much simpler, but perhaps more believable than the previously accepted facts of MacCreagh's early life. In the earlier *High Point Enterprise* article, we are told that MacCreagh

"began his official career when he went to Paris to study art at the age of seventeen. Seventeen-year-old students of art are as like to become plumbers as artists; this one did neither, but after being mixed up in a sabre duel and hastening abroad for a few months' sojourn in Germany, he went to Calcutta."

Here, the implication is that the duel happened in Paris, not Heidelberg. Although duelling was not as common in France as in Germany in the early 19th century, it did occur and so it is possible that this is where the duel really took place. However, in his 1921 US passport application, MacCreagh clearly states that he was in Germany from 1906-7 and France from 1907 to July 1909 "for education

---

[11] An article entitled *"All in One Lifetime"*.

purposes".[12] Given the conflicting versions, it is impossible to be precise about what actually happened and where. It is likely that either a forgetful MacCreagh confused the dates of his stays in France and Germany in his passport application, or the journalist writing the North Carolina newspaper article somehow transposed the events.

The *Mattoon Gazette*'s account written some five years later, is much shorter and is the first time in print that we learn of MacCreagh's Indiana origin. It tells us that MacCreagh

"ran away from home seventeen years after his birth in Perth, Indiana, and eventually he reached India."

Since both of these newspaper accounts were published many years after the actual events occurred, some confusion in the story is not altogether surprising. Although there is no record of MacCreagh having been an officially registered student at Heidelberg, it was possible at that time for unregistered students to use the university's facilities and even attend some lectures. So, this may well have been what MacCreagh was in fact doing whilst in Germany. Later, when he was living in New York, his profession is shown in a business directory as that of an artist which supports the likelihood of MacCreagh having attended art school in Paris. The fact that he apparently spoke both reasonable German and French gives further credence to his having spent time in both countries.

On the surface, the statement that MacCreagh ran away from home at seventeen makes some sense. It fits well with the play of subsequent events in his life and it somehow 'matches' his feisty, maverick personality. It also helps to explain the apparent absence of any contact or communication with his parents during his subsequent life. The only time MacCreagh mentions either of his parents is a very brief reference to his mother in *The Last*

---

[12] Curiously, MacCreagh fails to mention in his passport application his time overseas in England, Asia and Africa.

of Free Africa when he states that he is unable to remember her age.[13] He makes no direct references to his father in any of his writings. This all points to either an early estrangement from his parents or their demise while Gordon was still a child. The previously cited brief mention of his grandfather would seem to confirm that MacCreagh lived with him for some time as a youngster. Having run away from home, it seems possible that MacCreagh went to art school in Paris and having completed his studies after two years, moved on to Germany where he then attended university on a casual basis while he considered his long-term future. His few months in Germany, lying low after the duel, would have potentially given him the time and motivation to make contact with someone in India and to subsequently travel out there. Since the overall weight of evidence elsewhere indicates that he was involved in a duel, whether it actually occurred in Paris or Germany is ultimately of little importance.

Returning to the newspaper articles, sadly, neither of them provides any clue as to where MacCreagh was or what he was doing for his first seventeen years. Also, MacCreagh doesn't tell the reporters exactly where 'home' was when he ran away. The Mattoon Gazette version implies he was in the US at the time. If he was still in Perth, Indiana, presumably living with at least one of his parents, why did he later invent the story of attending boarding schools in Britain? Why did he choose to make the journey across the Atlantic to further his education in Paris or Germany? If MacCreagh did decide to run away while living in Indiana, destinations such as California, Canada or Mexico would have been easier, more natural choices for a young American man to strike out on his own, seeking adventure in a distant location. If he was in Perth, Indiana until he was seventeen, then what had George MacCreagh been doing all those years in such a

---

[13] See page 14 in the 1928 edition by The Century Co.

small, remote place? It's hard to believe that he was still researching the native Americans. We are also still left with the puzzling question of how a runaway MacCreagh was able to finance the expensive trip from Perth, Indiana to Britain, his stay in France and Germany of some two or three years as a student, plus his subsequent journey to India. The cost of tickets plus food and lodging along the way for these sea voyages would surely have been beyond MacCreagh's meagre teenage resources.

There are two other final pieces of information from the press to add to the jigsaw puzzle of Gordon MacCreagh's early life that are worth considering. The first is a biographical article about him that appeared many years later in the *Tampa Bay Times* on September 15th, 1947. The second is, perversely, his obituary, published in 1953 by the *St Petersburg Times* in Florida, which had then been MacCreagh's home for many years.

Taken together, these two articles contain some remarkable new information. Some of it supports the general thrust of previously published biographical accounts but other parts are curiously contradictory to the accepted 'facts' and again raise many questions. We are told for example that it was Gordon MacCreagh's Scottish grandfather who first emigrated to the US and that he fought in the Battle of Kennesaw Mountain in 1864 during the American civil war. If this information is correct, then it implies that Gordon's father might have been born in the USA, not Scotland. The statement in MacCreagh's 1921 passport application that his father, George, was born in Chicago, could therefore be true. Although this would help explain why there is no record of the birth of George MacCreagh in Scotland, the absence of his name in any US census remains puzzling.

The obituary confirms Perth, Indiana as Gordon MacCreagh's place of birth and goes on to state that his parents returned to Scotland with their son when he was four years old, though no reason is provided for this important decision. Clearly in this scenario, Gordon could not have attended school in Perth, Indiana as claimed in other earlier biographical sources. Although there is no mention of his grandfather accompanying them, he must have done so at some stage as all other evidence points to Gordon being raised by him in Scotland. Though why MacCreagh's grandfather chose to go back to become a church deacon after living in the USA for some thirty years is a mystery.

The MacCreagh's returned to what is described as the family seat – the remote island of Lismore in the Outer Hebrides, north of Oban in western Scotland. The fact that around half the island's population emigrated, mostly to North America, as a result of the Highland Clearances in the second half of the nineteenth century lends some credence to the statement about MacCreagh's grandfather going to the USA.[14] The use of the word 'seat' above, implies that the family had an established home on Lismore and were therefore of some standing locally. Yet, as already stated, there are no records of either any MacCreaghs living there or any other part of Scotland, nor of any transatlantic crossing during this time.

The press articles tell us that MacCreagh's father, George, worked as a noted collector of ethnological artefacts for the British Museum and other institutions. Also, that as a result of his job, George was frequently away from his home in Lismore, often for many months at a time. Indeed, we are told Gordon apparently accompanied his parents on two overseas trips, one of which took them to India and subsequently to Durban in South Africa where the young MacCreagh actually attended school for a while. This type of work would suggest that George must have been reasonably well-educated and knew something about the world and foreign

---

[14] The Highland Clearances involved the eviction of a significant number of tenants in the Scottish Highlands, mostly in the period 1750 to 1860. The clearances resulted from agricultural improvement, driven by the need for landlords to increase their income.

cultures. It also indicates a potentially decent, if sporadic, level of income, though hardly enough to have financed Gordon through the leading British boarding schools as claimed in other accounts. If this is what George MacCreagh was doing in the USA when Gordon was born, it is hard to imagine him operating from Perth, Indiana. Similarly, the island location of Lismore seems an unlikely base from which to carry out such employment. Since the British Museum has no record of a George MacCreagh (or any similar surname) working as one of their collectors, either as an employee or on a freelance basis, this version of events is highly suspect.

We are told that at the age of ten, Gordon returned to Britain with his mother where he remained, presumably largely in the care of his grandfather, until he was fifteen. At this point, it seems his parents moved to Europe, living initially in France and then Heidelberg in Germany where Gordon studied languages at Heidelberg and Gottingen universities. This contradicts the information in MacCreagh's passport application about going to Germany first and then France. A minor detail perhaps but still confusing enough to raise the question of why MacCreagh's story constantly varied. According to the obituary, the fencing duel that led to MacCreagh's flight to India, took place in Heidelberg. While this information gels with previous press report that MacCreagh went to art school in Paris first and then to university in Germany, it doesn't appear to fit with the Mattoon Gazette's account about running away from home at seventeen. If there was any running away, this most likely took place only after the fencing duel. It also seems odd that if MacCreagh's parents were truly in Heidelberg, why he felt unable to turn to them for help after the apparently fatal duel.

If the rather dis-jointed family life described above holds more than a grain of truth, it helps explain MacCreagh's apparent disconnect with his family once he left for India. He seems to have lost all contact with them and his father, mother and grandfather all disappeared from his life. Of course, the information contained in these two newspaper articles is only as good as the input which was largely based on anecdotal details supplied by MacCreagh himself, either directly or via what he told his wife, Helen, or his close friends. There is a distinct lack of hard factual evidence to support the claims made in either these two accounts or indeed, those in other published early-life biographies. There are two possible explanations for the apparent dearth of official data on MacCreagh's early life plus the confusing and conflicting details that have been passed down. The first relates to Gordon MacCreagh's surname.

It is based on a notable Scottish name, deriving from MacRaith, an ancient Gaelic male given name meaning son of grace or prosperity, from the Celtic word "rat" for luck or good fortune. The surname first appears on record in the early 13th Century and in more modern times has many spelling variations, ranging from McCray, McCrea, McCree and McCrie, to McCraw, McCreagh, McCraith and McRae. The spelling of MacCreagh is unusual and there is a noticeable lack of any contemporary Scottish records with this particular version of the name. When MacCreagh fled after his duel, afraid of legal proceedings, it may be that he modified the spelling of his last name in an attempt to disguise his identity or whereabouts. It would then explain why there is no mention of any MacCreagh in the relevant records for births, deaths or immigration. It could also be the reason behind the dubious birth certificate produced with his 1921 passport application. In this context, it is interesting that the Island of Lismore is the home of the Scottish Clan MacLea - a name not too dissimilar to MacCreagh. Without a definite surname to go on, it is impossible to be certain about MacCreagh's parents' origins and whether they actually ever lived in the USA or even if their son was born or lived there as a child.

Whether MacCreagh changed his name or not, why was he at best vague or imprecise and at times, contradictory or even lying about his early life? Why did MacCreagh apparently feel the need to reinvent himself when his early life had been so interesting anyway? Why did he take the risk of inventing the story about being born in Perth, Indiana and attending exclusive boarding schools in Britain? As he steadily became more famous in the 1920s and 1930s, the chances of being caught out in some way must have been a growing danger. Why did he apparently lie about something as basic as his date of birth? It's worth looking at the very first sentence in his 1933 autobiographical note from the *Argosy*, "I can't lie about my age because it's in *Who's Who*, and it's older than I like to believe." Was this his way of acknowledging that he had lied and that he was in fact not as old as the date of birth of 1886 given in *Who's Who* implied?

This brings us to the second possible explanation about what may have happened. It's an established fact that many of MacCreagh's contemporaries as writers for the *Adventure* and *Argosy* magazines are known to have embellished or lied about their past lives. The most infamous of these is probably the case of Talbot Mundy. His real name was William Gribben and he has some interesting parallels with MacCreagh, including the early death of his father. In a lengthy autobiographical account published in *Adventure's* "The Camp-Fire" section[15] in April 1919, Mundy described himself as an explorer and hunter, claimed to have worked in India as a foreign correspondent and to have fought in the Boer War in South Africa. He stated that when badly wounded there, he quickly recovered on learning that his grave was being dug and that he was expected to occupy it. However, the reality was completely different.

The young Mundy or Gribben was a scheming, untrustworthy scoundrel, with an extensive history as a swindler and a con-artist. Although he came from a well-

---

[15] A part of the magazine dedicated to readers' questions and discussions – see later comments in Chapter 5.

to-do English family, he turned into the "black sheep' of the family following his father's premature death. He was expelled from the famous public school of Rugby in his late teens and he then ran away, going first to Germany where he briefly worked as a driver towing circus vans, and later, to India and Africa. He was apparently named as co-respondent in at least one divorce and acted in such a promiscuous manner with native women while in Africa that he was arrested and escorted out of town on a number of occasions. When he emigrated to the USA, he became seriously involved in gambling until he was severely beaten by some of his fellow gamblers and ended up in hospital. While lying there recovering, he finally decided to mend his ways and spent the rest of his life making a living by writing fiction.

Perhaps even more relevant is the example of Captain Dingle, who we will hear more of later in the book. When MacCreagh arrived in New York in 1911, he shared a flat with Dingle for some time and they became close friends. His original full name was Albert Dingle but at some stage as a young man, he changed his first name to Aylward; he also lied about his birth date and modified some of his family background, including their origins. There are many more similar cases. With these mendacious and dissembling examples apparently all around him, it is entirely possible that MacCreagh decided to also re-invent himself and create a story that was different from and perhaps to his mind, even more interesting than the reality. When he arrived in India seeking employment, perhaps having changed his name, he was quite young and still only a teenager. So, it is entirely possible that he wanted to make out that he was older than he really was and therefore added a couple of years to his age. There is certainly ample evidence that in later life, MacCreagh regularly altered his date of birth so as to appear younger than he really was. This could have been the genesis of his new persona. Having once made this relatively minor change to the facts, he then stuck with it and gradually added to it.

In trying to piece together MacCreagh's early life, we have to recognise that he was an inveterate and accomplished story-teller, with a mischievous sense of humour. Like many famous people who are frequently interviewed, he probably tired of answering the same questions and modified or embellished the actual truth as he went along.[16] There is a very revealing sentence in the Mattoon Gazette article of 1928 referred to earlier that describes the interview with MacCreagh. The newspaper's reporter stated that on being asked a series of questions about his early life " MacCreagh becomes shy and reticent and forgets details."

It seems to me that MacCreagh effectively gradually invented his own back-story in much the same way that Lucas and Kasdan did for Indiana Jones. Only MacCreagh did this well before the term had even been thought of. In the end, we will probably never know the full truth about the early years of his life. Although the weight of previous, anecdotal information points to Perth, Indiana as MacCreagh's birthplace, the only evidence to support this is his rather suspicious 1921 passport application and his own subsequent statements. Many of the published biographical accounts of MacCreagh's early life do not appear to stand up to serious investigation and this raises many difficult-to-answer questions. The possible alternative version of having been born in Scotland of Scottish parents, emigrating to the USA in 1911 and becoming a naturalized citizen, as indicated by the US census of 1920, provides a more likely and credible backdrop to subsequent events. As we shall see later, there is additional first-hand evidence that lends a reasonable level of credence to this version. However, it's up to the reader to accept those elements of MacCreagh's early life that they feel are most plausible and reject those that seem less likely to be fact. Perhaps this is what MacCreagh, the inventive story-teller, really wanted people to do all along –

---

[16] There were dozens of articles in US newspapers and magazines written about MacCreagh following interviews during the 1920s and 1930s.

to take him as they found him and not judge him with preconceived notions based on specific details of his actual personal background.

Chapter Four

Asia and Africa

Despite all the questions regarding the who, what, when, where and how of MacCreagh's early life, there is no doubting that he eventually made his way to India, probably arriving sometime in late 1908 or 1909. Exactly what happened once he left Europe is unclear and even MacCreagh provided differing and conflicting versions of what he did and when. In his 1933 autobiographical note for the *Argosy*, he states that after the duel "everybody concerned laid low for a while" though he doesn't say where. In the 1900s, it would have taken time and money to make such a long journey from Germany. Although MacCreagh later stated that he worked his passage as a ship's steward, it is unlikely he would have taken the decision to travel to far-away India without some good reason. MacCreagh inferred that his decision to head to British India was taken while he was still studying in Europe. He claimed he had been in correspondence with a wealthy man in Calcutta who had promised MacCreagh an attractive salary if he came out to work for his shipping business. He went on to say that "the kind gentleman gave me the job (as an under steward). But at the end of a month when I asked him for some rupees, he said, Oh, yes, he'd give me them - as soon as I had learned the barge business and was of some use to him." MacCreagh added that as a result, he had a fight with his new boss's son-in-law and was fired.

Once again, we are left with more questions than answers. Who was this mysterious rich man of Calcutta and how did the pair first get in touch with each other? Why would this man offer the inexperienced, nineteen-year old MacCreagh a job with an attractive salary on his barges? Why was MacCreagh even considering such a move when he was still pursuing his education at university? Lastly, why did he agree to work as an under steward on an Indian river barge since it was hardly likely to be an interesting, or well-paid job? Sadly, we have no direct answers to these questions and can only speculate. Since MacCreagh wrote his brief autobiographical article almost twenty-five years after the event when he was at the height of his creative story-telling powers, it seems likely that he either forgot or once again, somewhat adapted the facts.

However, there are a couple of potential explanations. It is entirely possible that, while at art school or university, MacCreagh made friends with a fellow student whose wealthy father owned a barge business in Calcutta. Such contact could have sparked a fascination in MacCreagh's young mind with the exoticism of India – a place he described in his auto-biographical note as "where one shook the rupee trees and gathered wealth and glamour at the same time." There could then have been a conversation along the lines of "if you ever want to come out to India, there's a job waiting for you." If so, MacCreagh would have had this information in his mind when he had to make the rapid decision about what to do and where to go after the duel. India's exotic remoteness from the scene of his recent crime would have appealed. Also, he would have known that as it was ruled by the British, it would be easy for him to find employment there.

Alternatively, MacCreagh's mysterious correspondent may have been a relative which would explain how the initial contact arose and why an offer of employment was made. Calcutta was then the bustling capital and administrative centre of British India and the jewel in the crown of the British Empire. Many Scottish people had gone there to serve in the army, to work as missionaries or in business and trade. The Scots even had their own school, church and cemetery in the city and the city's financial, social and political nucleus, Dalhousie Square, was named after a Scotsman. Several men with versions of the surname of MacCreagh are recorded in local journals as being in Calcutta during the decades prior to Gordon

MacCreagh's arrival.[17] One of them might have been running a successful barge business and encouraged MacCreagh to come out and join him with stories of the riches to be made. The final alternative possibility is that the contact in India came from MacCreagh's grandfather. At the time, the various churches in Britain were extremely active with missionary work in India and as a church deacon, he may have put young MacCreagh in touch with someone there. The most obvious choice of a possible Calcutta employer was Yule & Co. Ltd., established by two Scots, Andrew and George Yule in 1863. They initially traded in commodities such as tea, jute and cotton but in the latter part of the 19th century, they also moved into shipping and set up several companies in Calcutta, including the Port Shipping Co. in 1906, which was then the largest bulk commodity lighterage on the Hooghly. It was a family business and employed the son of Andrew Yule – possibly the man that MacCreagh claimed to have had a fight with over his pay.

Whatever the explanation, there are several different versions of what happened, once MacCreagh arrived in India that conflict with his own story above. In one, we are told he found himself a job working for a lighterage company as the captain of a river barge working on the Hooghly River in West Bengal. Of course, MacCreagh had no experience of captaining a barge, so this seems rather unlikely. Another version of events is the short biographical account included in the 1985 reprint of MacCreagh's book *White Waters and Black*, written by Bob McKnight.[18] He was a good friend of MacCreagh and presumably, much of his information came first-hand from MacCreagh himself. In this, more plausible, version of the story, MacCreagh was working on a river barge that had had a crew of ten. He woke up one morning to find that

---

[17] E.g. the Calcutta Magazine and Monthly Register – see Volumes 10-12.

[18] This article was originally written by McKnight for the American Museum of Natural History in March 1960.

seven of them had died of bubonic plague during the night. The other three had apparently jumped overboard and swam to shore, an action that MacCreagh then sensibly followed. He eventually managed to make contact with his company, only to be told that "he would have to go back to the barge and await fumigation. This would have meant almost certain death." So instead, MacCreagh left and headed off in the opposite direction, looking for employment and adventure elsewhere.

Yet another alternative version of the story was that printed in the Illinois Mattoon Gazette in January 1933 following an interview with MacCreagh and just before his auto-biographical note in the Argosy appeared. Here, the barge had become stranded on a sand bar in the Hooghly River. During the night, the whole crew of ten Hindu coolies fell ill with bubonic plague and five were soon dead. The remaining five then stole the ship's lifeboat and slipped away in the darkness, leaving MacCreagh stranded alone on the barge with the five dead men. He remained abandoned there for several days, fortunately without catching the plague, until he was rescued and brought ashore. Interestingly, none of these accounts mentions MacCreagh's issue with payment or the fight with the barge owner's son-in-law. Like most good stories, especially ones told by creative writers like MacCreagh, details change with the telling and some events are embellished for the audience. Regardless of whether by invitation or through happenstance, there seems no doubt that the young MacCreagh worked for a short while in some capacity on a river barge on the Hooghly. As for the rest, I leave it to the reader to decide which version lies closest to the truth.

Hooghly River Calcutta c.1895

Old Court House Street, Calcutta c.1900

Whatever the reason for leaving Calcutta, MacCreagh was young and he no doubt did what most young men would have done to finance their travels – he took odd jobs along the way. According to MacCreagh's own account, our erstwhile barge captain then caught a train north to Darjeeling where he found a job for a while as an assistant overseer at a tea plantation. In the latter part of the 19th century, the West Bengal region of India had overtaken China to become the world's leader in the production of quality tea and there were hundreds of tea estates in the area. However, MacCreagh soon found this work far too boring and so he set up a side-line collecting specimens of the rare and spectacular butterflies, beetles and orchids found in the foothills of the Himalayan region.[19] From this small beginning, he developed an interest in exotic fauna and flora that would endure for the rest of his life. For a while, he sold these on to museum collectors, including the British Museum but he later graduated to collecting for himself, generating sufficient money to help finance his subsequent travels.

Whilst in this part of the world, MacCreagh decided to venture further north into Nepal and Tibet, though whether this was in search of rare species, gemstones or simply for the fun of exploring and visiting out-of-the-way places is unclear. Whatever the reason, it must have been an exhilarating experience and a challenge both physically and mentally. To reach Tibet, MacCreagh crossed over the Jelep La pass which lies at 14,000 feet on the ancient trading route connecting Sikkim in India to Lhasa, the Tibetan capital. He made this remarkable journey alone and with not much more than a simple rucksack on his back.[20] Sturdy legs and strong lungs were needed to make such a high-altitude trek – MacCreagh clearly had both.

---

[19] MacCreagh developed a life-long interest in orchids – see comments at the start of Chapter 11.

[20] In his later book, *White Waters and Black*, he describes the pass as the most evil place which he had the misfortune to traverse.

The area was an extremely dangerous place to travel through in the early part of the 20th century, especially by oneself. Just a few years earlier in 1904, the British had sent an expeditionary force through the pass and on to Lhasa to counter the growing Russian influence in Tibet. This expedition was met by hostile Tibetan forces and although they were defeated by the British, the area remained very volatile. At the time of MacCreagh's trek, the Chinese were also threatening to invade Tibet, something they did just a few months later, forcing the Dalai Lama to flee through the pass to India.

While in India, MacCreagh also developed another profitable venture, gathering and selling trophies to hunters who couldn't catch or shoot their own game. From these small beginnings, MacCreagh soon found that there was a ready market for much larger creatures. He began supplying live animals to dealers in England, especially to a company called Jamrachs Menagerie in London. The Jamrach Company was founded by Charles Jamrach, a German émigré who moved to London in 1840 and established what would become the world's leading dealer in wildlife, birds and shells. He also owned an exotic pet store in East London that at the time was the largest such shop in the world. He bred, imported and exported animals that were sold to noblemen, zoos, menageries and circus owners around the world. By the time of MacCreagh's involvement with the business, it was being run by Jamrach's two sons. Their retail prices for animals were high – lions sold for in excess of £100 and tigers for £300 but for some reason, leopards were relatively cheap at £20. This was clearly the kind of work that appealed to the adventurous nature in MacCreagh and he took to it with gusto.

He would later claim that at one stage he bought himself an old jalopy that he used to collect his wild animals and transport them to the railhead. On one occasion, the Bengal tiger trussed up on the back seat apparently partially freed itself and started to attack MacCreagh. He had to immediately pull over and leap out of the car until he could secure the wild creature once more. He also claimed that he drove right across India in his jalopy, often travelling along railroad tracks when there were no roads. The work proved sufficiently profitable that MacCreagh decided to move on, first to Borneo and then over to Malaya where there was a more plentiful supply of the animals in demand such as leopards or tigers and MacCreagh's particular speciality – big snakes and orangutans. In Malaya, MacCreagh found employment with the Kedah State government catching and caring for wild elephants for export. He apparently lost his job when he allowed a marauding tiger to get too close to one of the elephants in his care – with fatal consequences for the elephant. So, when MacCreagh later vividly describes in his pulp fiction adventure stories how the hero of the story is attacked by wild animals, we know he is speaking from real-life experience.

After a few months, he moved on again, initially returning to Calcutta from where he then set off to explore the wild, back-country along the north-eastern frontier of Burma and China. He travelled with another young British man that he had met earlier in either Calcutta or Darjeeling. They travelled first to Mandalay but at that time, there was no rail connection between India and Burma so the pair presumably walked or covered the two hundred-mile journey by mule. From there, the two young travellers took a train north to the town of Myitkyina before heading south again to Lashio, the regional capital of the remote Shan State region.[21] At some stage during this trip, whether alone or with his British friend, MacCreagh also ventured into China, visiting the province of Yunnan.[22] Incredibly, during their Burmese travels, the pair began operating an open-air movie for the natives. Since the natives had no cash, it should have been clear to them that

---

[21] The Shan States were then separate from Burma but then under British administrative control.
[22] This region would be visited just a few years later by Roy Chapman Andrews, one of MacCreagh's rival candidates as the real-life Indiana Jones.

this would be no money-making venture. Apparently, MacCreagh and his partner were obliged to accept chickens and pigeons as payment for tickets by the locals. According to the Mattoon Gazette "all went well, until a native handed in a goat at the box office. MacCreagh thought a bit, then gave him a ticket and five chickens in change." [23] Whether the story is completely true or not, it illustrates both MacCreagh's perseverance in the face of difficulty as well as his enduring sense of humour that we will encounter time and again as his adventures continue.

The open-air cinema was probably just an amusing diversion for MacCreagh in Burma while he increasingly developed his more profitable endeavours in trapping and hunting big game. From Lashio, the two young men travelled in comparative luxury by the little twenty-two inch-gauge railway back down to Mandalay. It was around this time that MacCreagh started to write his first book *Big Game in the Shan States*, presumably based on his recent experiences along the Burma-China frontier. As McKnight says, "The fact that he knew nothing about writing was no deterrent." [24] Sadly, no extant copy seems to have survived and the book was most likely a limited edition run, printed in Calcutta, possibly by Thacker, Spink and Company, Calcutta's leading printer of the time. The Indiana University publication Indiana Authors and Their Books gives 1909 as the date of publication but since MacCreagh had probably only arrived in India that year, this date looks a little early. Whatever the date of publication, this early evidence of his writing skills at such a young age was quite an achievement.

---

[23] Ibid
[24] Ibid

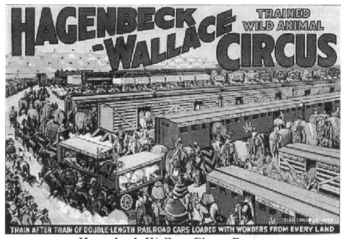
Hagenback-Wallace Circus Poster

Unloading Live Animals, London Docks c.1890

While in India and south-east Asia, it appears MacCreagh took a growing interest in the local tribes-people and their habits, customs and folklore. His interest in and resultant knowledge of the relatively primitive peoples and tribes that he met during his travels was a trait that would stay with him during all his subsequent travels. He also seems to have had the marvellous ability to pick up the basics of their languages and dialects, regardless of whether it was those of the Burmese tribesmen, the Amazonian Indians or the natives of Africa. This ability to converse with local people allowed him to not only get closer to them than many other contemporary writers but it also provided him with a rich vein of background material for many of his subsequent stories.

At some stage, MacCreagh claimed he also worked for a ruby mine in Mogok, Burma where he found a fine stone which, much to his dismay, was stolen from him at gunpoint one night in the camp. Eventually, after yet more adventures, the bug to move on and explore more of the world once again took hold and MacCreagh left Burma and travelled to Siam (Thailand) and then to the Dutch East Indies (now Indonesia). He financed his travels this time by working for a short while as a surveyor – a skill that would prove useful in his later, more serious journeys in South America and Ethiopia. He remained in south-east Asia long enough to save sufficient money to pay for his next travel adventure – a journey to Africa. It is impossible to know the motivation for MacCreagh's constant drive for new adventures and to explore. If his father was really a historian and naturalist who regularly travelled overseas, the compulsion may have been inherited. Similarly, MacCreagh's interest in other peoples and cultures could have derived from the same source in a similar way to Indiana Jones' enquiring intellect deriving from his father.

From Jakarta, MacCreagh now made his way by ship across the Indian Ocean to the port of Mombasa, Kenya in what was then British East Africa. From there, he travelled inland by train to Nairobi, the newly-founded administrative centre of the growing British colony. Once there, it seems he was soon able to talk himself into a contract to supply wild animals for the Hagenback-Wallace Circus. In the early part of the 20th century, the Circus was at its peak, travelling across the USA, mostly by railway in a dedicated convoy of wagons. It was the second-largest circus in America, competing with those run by the Ringling Brothers and Barnum and Bailey. Coincidentally, the Hagenback-Wallace Circus was based in Peru, Indiana, a little further east along the Wabash Valley from MacCreagh's possible birthplace. It was a risky task to take on – he had only just arrived in Kenya and knew little of the country and had no structure in place to deliver such a contract. However, to quote from Bob McKnight, "MacCreagh had that amazing quality of being ready and willing to tackle anything, regardless of whether he knew anything about it or not, or how dangerous it was." [25]

MacCreagh initially based himself at the Norfolk Hotel in central Nairobi.[26] While staying at the Norfolk, he made friends with the American, Alfred Klein, who was an experienced big-game hunter and would later become a well-known safari expedition leader. Klein spent fourteen years in Africa, shooting around one hundred lions as well as hundreds of other wild animals. He later had the audacity to claim there was little big game left to shoot in the area because the natives had killed it all.[27] The two men shared a common interest in big-game and no doubt, spent time together exchanging their different experiences of hunting wild animals. It may well have been Klein who helped MacCreagh to become established in his new

---

[25] Ibid

[26] The Norfolk became the social focal point for the whole of British East Africa, especially as the preferred watering-hole of the notorious Happy Valley set of dissolute aristocratic settlers.

[27] Article in *Popular Science*, December 1924. Klein's Camp, an up-market safari resort overlooking the Serengeti Plain is named after him.

enterprise of supplying circus animals, either working with him or by simply providing advice, contacts and local wild-life knowledge. It is probably no coincidence that a hunter called Klein features as a character in MacCreagh's 1930 story, *The Slave Runner* which is set in Kenya.[28]

MacCreagh spent several months freely roaming the vast east African plains and found it relatively easy to collect the wild animals he needed. He also acquired something else during his travels – a detailed knowledge of the way of life of both local tribes and the European settlers. He would later use this knowledge to good effect in the plots and characters of his many adventure stories that were set in Africa. As it turned out, MacCreagh was so successful in his new enterprise that he soon found himself with more animals than the circus required. Unable to find alternative customers for this surplus, he ended up spending almost all the profits he had made on caring for these unwanted animals.

The above accounts of MacCreagh's experiences as a young man in Asia and Africa demonstrate his adaptability and bravado plus the ability to make and develop business contacts. He was clearly able and willing to take on almost any kind of employment and as we will see in his subsequent life, he never let a lack of experience hold him back. Once again virtually broke, MacCreagh decided it was time to move on once more. He had just enough money in his back pocket to return to Europe and spend some time traveling around Germany. It is possible that he may have also visited his parents if they were still living there. He evidently felt sufficient time had passed since his university duelling escapade so that he was in no danger of being hauled up before a German court.

MacCreagh was now at something of a crossroads. He was twenty-one, had little money, no job and no real qualifications. Although he had immensely broadened his

---

[28] Klein is referred to as being "known throughout East Africa to have killed some hundred and fifty lions".

knowledge of the world through extensive travel and a fascinating array of short-term jobs, he had no real immediate prospect of any worthwhile career. Once again, he elected not to return to his home in Britain or the USA and try his luck there but instead he decided to go back to India, arriving once more in Calcutta. It seems odd that he chose to return to the city. His previous stay there had not exactly been a memorable experience. It was a place he would later describe as being "one hundred miles up the most treacherous river in the world . . . a site of accumulated sewage and silt." The fact that he did go back suggests he probably had friends and contacts there.

However, as so often with MacCreagh, we again find that there are different versions of exactly what he now did and for how long. According to the primary biographical sources quoted earlier, he soon found work with the intelligence service of the British Post Office, a job that he went on to hold for five years which if true, was a remarkably long period for him. During this time of the British Raj rule, although the Post Office in India operated part of that country's internal intelligence gathering service, it was primarily an information compiling and collating function. It had hardly any independent channels for conducting pro-active intelligence gathering. Under the Indian Post Office Act, it carried out the surveillance of telegrams and printed materials plus the monitoring and interception of imported goods from seditious pamphlets to guns and ammunition. Any such work would have been largely paper-pushing, hardly the kind of job to attract and retain a young man like MacCreagh for five years.

At the time of MacCreagh's arrival, the particular concerns of the Imperial Government were the dual threats of incursion by Tsarist Russia in the north and sedition or local uprisings in Bengal. In order to boost the numbers of skilled personnel, the Post Office had recently mounted a recruiting drive, offering increased salaries and often an extra allowance for local accommodation to any man or woman that joined. Since MacCreagh was once again basically penniless, it may well be that he joined the Indian Post Office on this basis; the offer of a decent, regular paycheck and subsidised accommodation would have been very attractive. Normally, such sensitive positions were reserved for British or Anglo-Indian citizens so assuming MacCreagh was born in Scotland, this would not have been a problem. However, if he was born in Perth, Indiana, as he later claimed, he must have emphasised his British schooling and put on his best English accent for the interview process.

However, if we accept as correct the information given by MacCreagh in his 1921 US passport application discussed previously, then he didn't leave Europe for India until late 1908 or 1909 at the earliest. Since the 1920 US census has MacCreagh recorded as arriving in New York in 1911, that only gives a period of some two or three years in which to have accomplished all his travels in India, South East Asia and Africa plus his second stay in India working for the British Post Office. So, clearly, MacCreagh didn't work for the British Post Office in India for five years; logic suggests a time-frame of just a few months as being more likely. Also, although the statement that MacCreagh worked for the intelligence service within the Post Office sounds somewhat glamorous and impressive, it is highly suspect. In his interview with the Mattoon Gazette in 1928, MacCreagh let slip that he was actually employed in a basic printing job within the Post Office itself. Despite the relatively low level of his job, it would have given him a badly-needed income and it seems he was happy and settled into colonial life, at least for a while. But the mundane nature of the work combined with MacCreagh's restless and energetic nature caused him to yet again cast around for a new activity.

He eventually turned his hand once more to writing, presumably while still working for the Post Office. MacCreagh initially dabbled with a collection of short stories but it was never published. From this, he progressed to writing a two-act play that he cast with native Hindu actors

East 14th Street, Manhattan 1918

New York Roof Top Theatre

and a Hindu princess in the lead role. It met with some success in local productions and was subsequently seen by a New York producer named Michael B. Leavitt who happened to be on a visit to India. He had made a name for himself in the American theatre industry by importing interesting new acts to tour the country. He also had a taste for sensationalism and no doubt saw an opportunity with MacCreagh's exotic play. He encouraged MacCreagh to bring his Hindu troupe to New York at Leavitt's expense. Enticed by Leavitt's attractive offer and the possibility of fame and fortune on Broadway, MacCreagh resigned from the Post Office and sailed with his troupe to New York. The play duly opened at the Amsterdam Roof, an open-air theatre on Broadway which, as the name suggests was situated on the roof of the building. It was part of the New Amsterdam Theatre in the main building below which was completed in 1903 in Art Nouveau style and cost $1.5 million to construct (equivalent to over $300 million today). At the time, it was the largest theatre in the city. The Roof Garden, where the more risqué productions were presented, and which no longer exists, was added on top of the building in 1904. At this time, New York City was literally peppered with similar rooftop theatres and the entertainment they provided was very popular with New Yorkers. Despite some initial success, the play was soon shut down by the New York authorities because it contained too much nudity, much to the annoyance of Leavitt and the disappointment of MacCreagh. The Hindu princess and many of the players were shipped back to India but MacCreagh was once more penniless and stranded in New York without a job.

Sadly, no copy of the play has survived. But if the story of bringing an exotic Indian Hindu princess all the way to New York to star in a risqué vaudeville production sounds too far-fetched to be true, MacCreagh supports the likely reality of the tale in two other sources. Firstly, in his semi-autobiographical, short story, Jehannum Smith, published in 1919, he describes how the eponymous hero is persuaded to free a dancing girl from a temple so that she can escape and be exhibited in the States. He gets into the temple, fights the priests, confronts a holy ape and escapes with the dancing girl, leaving a trail of destruction behind. A story-line that could easily have been an exciting scene from an Indiana Jones film. Secondly, in his interview with the Mattoon Gazette in 1928, he repeats much the same detail, so the apparently far-fetched story would seem to be largely correct.

Chapter Five

New York, New York

When Gordon MacCreagh arrived in the USA from India in 1911, either as a new immigrant or as a returning son, the country was undergoing rapid and significant change. The population of New York, at close to five million, had almost doubled during the past twenty years. The arrival of electrification, the telephone and the internal combustion engine had transformed not only the economy but also the everyday lives of most Americans – at least in the cities. The USA had become the world's largest industrial producer, generating as much manufactured output as France, Germany, and the United Kingdom combined. However, the skills required by the booming coal, steel and industrial manufacturing sectors were not ones possessed by MacCreagh. As the author would later write about the lead character in the semi-autobiographical story, *The High Flier*, "I have no experience in manufacture.'" According to the interview he gave the *Mattoon Gazette* in 1928, "MacCreagh found his first years in the civilized metropolis more difficult than any he had experienced." Never a man to give up easily, as we have seen, MacCreagh began to cast around for new opportunities. Fortunately for him, high employment, reduced working hours and increased disposable incomes had ushered in a social change that was to prove very beneficial.[29] That change was the significant growth in popularity of adventure fiction magazines, primarily aimed at an adult male audience.

---

[29] There was one other change that occurred in 1913 that was not so beneficial – the introduction of federal income tax. Prior to that date, the US had no federal income tax, no central bank, no social security taxes and no general sales taxes. Of course, this would not have worried the penniless MacCreagh.

At that time, there were several well-established, popular magazines with national circulations. Of these, The Argosy is generally regarded as the first pulp magazine and was started in December 1882 as a boy's adventure story paper, edited by Frank Munsey. It ran as a weekly until April 1894 when it became a monthly with a marked shift in content aimed towards a more adult readership, before returning again to a weekly format in 1917. At first, it combined both general articles together with adventure fiction stories but by the turn of the 20th century, it had changed to an all-fiction format and enjoyed a swift rise in popularity. Its policy of maintaining four or five serialised stories running in each issue helped to keep readers loyal and drive up its circulation. The magazine would eventually publish over 2500 issues in its various incarnations.

Competing with The Argosy was the All-Story Magazine which had first appeared in 1905, also founded by Frank Munsey. Under its ambitious editor, Bob Davis, the magazine gained an envied reputation for discovering budding new literary talent. In the All-Story's fifteen-year life, it published stories by authors such as Edgar Rice Burroughs (Tarzan of the Apes),[30] Max Brand (who later wrote the first Dr Kildare in 1938), John Buchan (The Thirty-Nine Steps in 1915), Zane Grey (Western stories), C.S. Forester (Horatio Hornblower stories) and Johnston McCulley (the Zorro tales). In 1920, the Argosy and All-Story merged to become The Argosy-Allstory Weekly, turning the new publication into the world's most widely circulated action magazine. The Argosy-Allstory Weekly continued under that title until 1929, when it returned to the original name of Argosy.

The third leading magazine of this genre was *Adventure* which was first published in 1910, initially printed fortnightly and later as a monthly. It became one of the

---

[30] Remarkably, a copy of the 1912 edition featuring the very first Tarzan story sold at auction in 2006 for $59,750.

most profitable and critically acclaimed of all the American pulp magazines, selling for 25c compared with its major rival, The Argosy at only 10c. One of the most popular parts of the magazine was a section called The Camp-Fire. This featured editorial comments, background information by the authors to their stories and discussions by the readers. Due to the positive reaction of its readers to this section, Adventure established a number of Camp-Fire Stations – physical locations where the readers of Adventure could meet up with each other. By 1924, there were Camp-Fire Stations operating across the US and in several other countries, including Britain, Australia, Egypt and Cuba. The set-up was rather like a modern-day internet members' forum.

As the pulp market developed and expanded, publishers vied with each other by introducing a wide variety of new titles and spin-offs in both the USA and overseas. At its peak, during the 1930s, there were over 150 separate titles and the most successful sold several hundred thousand copies each week. Adventure magazine became internationally famous, with a fanatically devoted following that numbered in the hundreds of thousands. In 1935 it was hailed as "The No. 1 Pulp" by Time magazine and in 1942, its publisher also acquired Argosy. However, when the market dramatically declined in the 1960s and 1970s, Adventure would become a dying embarrassment, printing grainy black and white pictures of semi-nude women and attempting to survive as a men's magazine in a market increasingly dominated by Playboy.

It was against the background of pulp fiction's significant, early growth that MacCreagh began seeking some form of employment. Several months after he arrived in New York, he and his pet python, Billy, started sharing an apartment with an Englishman called Captain A. E. Dingle. He was a few years older than MacCreagh and since the age of fourteen, he had spent most of his life as a mariner. Like MacCreagh, Dingle had travelled extensively and during his many years at sea, he had been shipwrecked a remarkable five times.[31] The two men shared a love of travel and adventure, a common heritage, had similar stories to recount and evidently, got on well together.

Dingle's first shipwreck occurred in 1893 when he was on a salvage ship that set out from Mahé in the Seychelles seeking gold that had gone down with the passenger ship *Strathmore*[32] off the French-owned Crozet Islands in the southern Indian Ocean. They found the sunken wreck, and its strongbox, but were unable to remove it. Eventually, they were driven off by gales and during the return voyage, their vessel was wrecked off another small island. The crew reached the island and survived for twelve weeks by eating rabbit, goat and fish until they were rescued by a passing French ship. On the plus side, while stranded on the island they did some exploring and found gold from a buried 1870s wreck. Dingle's final command was also shipwrecked. He was returning to Britain from Australia in 1911 on the *Mara* with a load of frozen lamb. Seepage from the ship's refrigerators started to accumulate in the bilge, so he ordered the sea cocks to be opened to let in sea water to reduce the smell. A few minutes later, the ship exploded under him – the sea water had somehow reached the boilers. He was thrown from the ship and was fortunately rescued by an American schooner. As a result of this incident, he lost his master's papers and without them, could not get other jobs at sea.

Once back home, a friend convinced Dingle that New York had more opportunities than England, so he sold his home, left the cash with his family and landed in New York in early 1912 after working his way across the Atlantic as a waiter on a steamer (the same method MacCreagh had used to initially reach India). In order to earn a living, Dingle tried a variety of work until he finally

---

[31] In 1942, Dingle appeared on the BBC radio programme *Desert Island Discs* and was probably the show's only experienced, genuine castaway.

[32] Wrecked in fog off the Crozet Islands in July 1875

got a decent office job. In his spare time, he started writing – he certainly had a wealth of real-life adventure to draw on. As luck would have it, he was invited to a dinner for adventurers in New York and over dinner, he related a story of his seafaring days.[33] One of the other diners was Arthur S. Hoffman, who had only recently taken over as editor of the *Adventure* magazine. He asked Dingle to write up his story and send it to the magazine. After some weeks of difficulties (Dingle didn't own a typewriter at first), he finally managed to type up the article, *Blind Luck on St. Paul*, and send it in to Hoffman. Dingle's story appeared in the January 1913 issue and he was paid around fifty dollars for his work.

It was at this time that Dingle and MacCreagh (plus Billy the snake) started rooming together and Dingle's success with the *Adventure* magazine spurred MacCreagh on in his literary endeavours. With the success of his play, albeit short-lived, MacCreagh rightly felt that he could develop his literary talent further. His travels in India, south-east Asia and Africa provided him with both a unique treasure trove of potential ideas for adventure stories and a personal knowledge of the local people to help create characters and story-lines. So, MacCreagh stayed on in New York with Captain Dingle and began writing seriously. Sadly, after his initial story, Dingle hit a dry spell and for the rest of the year, he didn't sell a single story while MacCreagh had yet to have anything published. At one point, their funds were so low that they reputedly lived for a week on a fifteen-cent tin of boiled, unsalted beans. After that hungry week, they decided that they'd had enough of beans, and went out to try to earn some money. A local boxing club was offering five dollars to each participant in a competition. Each participant had

---

[33] This was the founding meeting of the Adventurers Club in New York of which both Dingle and MacCreagh were members for many years. There were then several such clubs in New York, including the Ends of the Earth Club of which Mark Twain was a member.

to take part in a fight, otherwise they'd be thrown out. So, the penniless pair went along to the club and signed up to fight each other. When they went into the ring, they were both so scared they'd be sent home without their five dollars that they pulled no punches. That evening, they went back bruised and sore but richer, with ten dollars between them.

Apparently, in another attempt to earn some money, MacCreagh even joined a New York music band as a bagpiper for a while, a skill that we assume was acquired during his schooldays in Scotland. Eventually, MacCreagh's persistence paid off and towards the end of 1912, he sold his first story to Adventure magazine – *The Jade Hunters*. Having set the scene in Burma, MacCreagh introduces the reader to a trio of characters who are on an elephant hunt but through a combination of circumstances, become diverted into a search for jade. We know little about two of them but the third, called Snake O'Shane, is described as a "Scotch-Irishman of medium height and powerful build, of whom it was said that he tracked tigers on foot, and who had been a miner, hunter, prospector, wild animal catcher, orchid seeker, and half a dozen other things during the twenty-four years of his eventful life." It is clear that this early MacCreagh tale is based on his own experiences in Burma just a few years before and the description of O'Shane, together with his nickname (MacCreagh liked snakes), indicates that he is a caricature of the author himself. Intriguingly, the age of the fictional O'Shane is exactly the age of MacCreagh in real life at the time of writing this story. Overall, the description of this story's lead character further supports the earlier-stated theory that MacCreagh was probably born in Scotland and was therefore British, not American.

Much the same conclusion could be drawn based on the leading characters in his five Neil MacNeil stories published in 1916 or the adventures of *Jehannum Smith* which first appeared in 1919. Set in India, these latter stories feature the eponymous hero, a rather idiosyncratic character, in his trials and tribulations dealing with life under the British Raj. Smith is clearly British and since MacCreagh later inferred that these stories were semi-autobiographical, it lends further weight to the theory that he too was British-born. When MacCreagh subsequently decided to invent his back-story about being born in Perth, Indiana, his later adventure stories began to feature American heroes instead, such as the hard-boiled Westerman in *The Crawling Script* story from 1923 and the wily, trouble-shooting King in the *Kingi Bwani* series which started to appear in the 1930s.

Around the same time as his very first story was published, MacCreagh joined the Adventurers Club in New York, attending its founding meeting on December 7, 1912. The club was a private men's club established by Arthur Hoffman, editor of the *Adventure* magazine. Also attending that first meeting were Captain Dingle plus several other pulp fiction writers. The membership soon grew to encompass a very broad cross-section of people including politicians, former military officers, newspaper correspondents and explorers. MacCreagh's membership of this club over many years almost certainly influenced his life in several ways. Mixing regularly (the club met monthly) with other adventure story authors probably helped MacCreagh in his efforts to develop his writing career. However, as previously implied, it may well have been the influence of the more nefarious of these authors that persuaded MacCreagh to re-write his back-story. In later years, the widening club membership probably also provided MacCreagh with useful contacts for some of his future endeavours.

MacCreagh's first paid story was soon followed by two more in 1913 - *The Brass Idol* was published in the October edition and *The Getting of Boh Na-Ghee* appeared in November. None of these early stories could be called great works of fiction but they were well received by the magazine's readers and their style and format set the basic formula that MacCreagh's adventure stories would follow for the next forty years. Three more stories were published in Adventure in 1914, followed by six in 1915. In August that same year, MacCreagh also had his first story published by the Argosy magazine and he was now well on his way as a recognized author. Soon, MacCreagh was selling his adventure fiction to a wide variety of newspapers and magazines across the country and abroad. He sold his first story in Britain, Everett, Commissioner of Justice, also set in Burma, which was published by Colliers in January 1915 as part of their collection of Greatest Short Stories. Captain Dingle's mojo also finally returned and he too had a number of stories and books published, mostly with a nautical theme, both in the USA and Britain.

We now encounter further enigmas or contradictions in the MacCreagh story. When the USA entered World War I on the side of the Allies in 1917, MacCreagh tells us in his *Argosy* autobiographical note of 1933 that he returned home (from Asia) and joined the US Navy, where "he lost a lot of time", apparently spending most of it at a Navy training station.[34] These are odd remarks for him to have made. Firstly, as we have already seen, he had left India in 1911 and was already living in New York in 1917 which was when he joined the US armed forces in the war against Germany. Secondly, in his book, *White Waters and Black*, published in 1927, MacCreagh described himself as "a one-time U. S. Navy first-class carpenter's mate ... and an ex-navy man can do anything." This statement implies his war service in the US Navy was at a relatively mundane level.

---

[34] Ibid

Adventure Magazine Front Cover, 1915

Adventure Magazine Front cover, 1918

However, in contrast to this version of his war service, some sources, including his Who's Who entry state that MacCreagh served as a pilot in the US Air Force, not the Navy. The MacCreagh account in Gale's Literary Resource also confirms the air force version and adds that he served in Europe for a year, though there is no corroboration of this assertion elsewhere. The apparent contradiction in these two different versions of MacCreagh's war service is probably due to the fact that there was no actual US Air Force in World War I. When the USA joined the War, its aircraft were operated either by the Army or the Navy. Thus, it seems likely that MacCreagh did indeed join the US Navy but at some stage, served in one of its aeronautical detachments.[35] Joining the US Navy's air service in 1917 and learning to fly would have been just the kind of adventure that MacCreagh would have sought and seems more likely than spending his time as a carpenter's mate. It also coincidentally matches the skill-set of Indiana Jones who occasionally takes the controls of an aeroplane in his films.

This latter version of events is confirmed by a supporting letter to MacCreagh's 1921 passport application from the New York Police Department. The letter, signed by the head of the New York Police Reserve Air Service indicates that MacCreagh was a lieutenant in his unit. The NYPD Air Service had been initially founded in January 1918 in response to the threats posed by German U-boats and sea-planes to the eastern coast of the United States. After the War, a group of around twenty veteran aviators were signed up as reservists to supplement the newly-formed police air service. They flew Curtiss bi-planes on patrol over the coast and city of New York and it would seem that at some stage, MacCreagh joined them. He would not have been able to do this without good

---

[35] The US Navy's air detachment was actually the first American military unit to be sent to Europe, arriving in France in June 1917. Whether MacCreagh served in this or a subsequent unit in Europe, as indicated by Gale's, has been impossible to verify.

previous flying experience. Also, several of MacCreagh's stories have an aeronautical theme such as *The Flying Chance*, published in December 1917 and *The High Flier* from March 1920. The detail that they contain about flying and aircraft seems to add further weight to the likelihood of his having served in the US Navy's air force at some stage.[36]

Intriguingly, in his *Argosy* autobiography, he tells us it was while in the Navy that he "met a god called discipline." It's a curious, yet interesting remark for him to have made. He clearly had something important on his mind to have included this phrase in what was a very brief, one-page biography. Was he referring to the sudden shock of encountering a strongly disciplined military environment in the Navy? While it is reasonable to conclude that MacCreagh's recent life had been without much structure and self-centred, even hedonistic, he would certainly have encountered discipline during his school days. If he meant that his time in the Navy taught him the importance of having a strong focus and structure in his life, there is little evidence to show that he adopted a significantly more disciplined life-style after the War. As we have seen, he was for quite some time unsure of whether he was an artist, photographer or writer.

Coincidentally, his former room-mate, Captain Dingle, also served in a US Navy training school during World War I, though with his maritime background, this made sense. He continued to be accident-prone however, and towards the end of the War, he set out from New York in a twenty-eight-foot sloop, intending to sail single-handed to Bermuda, He took his pet Airedale dog along on the boat for company. It was a hazardous trip to try to make. German submarines were still very active off the eastern coast and the hurricane season was due. Although he managed to avoid the Germans, after seventeen days at

---

[36] The Massillon, Ohio *Evening Independent* of December 13th 1935 also refers to MacCreagh being a pilot in the US Airforce in WW1.

sea, he was struck by the worst hurricane in thirty years when just twenty miles off Bermuda. Dingle was blown off course and almost wrecked. With a single tattered sail, it took him a further nine days to struggle back to Bermuda. For the final five days, he was without water and forced to feed his dog on liquid from the dwindling store of canned goods aboard.

It seems that MacCreagh continued to write while he was still serving in the US Armed Forces, with several stories published in 1917 and 1918. After the War, his output picked up for a while with stories appearing in *Adventure*, *Argosy* and *Munsey's*. He also wrote his first novel, a three-part serial, *The Naked Men of Naga*, which appeared during April and May 1920. After this burst of activity, MacCreagh only published one story in 1921. Unusually, this was a costume drama set in France with the tile of *A Good Sword and a Good Horse* and it was one of his few adventure stories which didn't have a jungle or expeditionary background. The publication of this story in *Adventure* marked the first time that MacCreagh appeared as a featured writer on the front cover. A clear sign of his growing popularity among the magazine's readers and in all probability, it also meant an increase in his author's fees.

With the growing success of his writing, MacCreagh's financial position dramatically improved and he was apparently able to make at least one return visit to his old haunts in Asia. We don't know whether this was because of 'itchy feet', the desire for more adventure or simply a bit of research for his subsequent publications. Given MacCreagh's character, it was probably a combination of all three. MacCreagh's creative story-writing ran in fits and spurts throughout his lifetime. In some years he wrote only two or three and in others, his output occasionally increased to six or seven. Some of his stories were quite short while others were either serials or book-length novels. In quite a few years, he published nothing at all.

Basically, it seems he wrote only when he had the urge and needed money to tide him over between his travels. As the years passed, once he had sold enough stories to the pulp fiction magazines, he was either overcome by the urge to travel or took part in some form of overseas expeditionary activity. We don't exactly know the details of all the travels and expeditions he went on during his life nor what jobs he might have held. As Peter Ruber said in his account of MacCreagh's life, "One leans to the opinion that a chronological reading of his adventure fiction may provide clues as to where he was or had been recently." [37]

The lack of output from MacCreagh's typewriter in the early 1920s may have been because he was still exploring different potential avenues in terms of his future career. While he listed himself as an author in the New York city business directory for 1920, by the time the 1922 edition was published, his profession had changed to that of an artist, living on East 14th Street. Yet in the US census of 1920, he had described his profession as that of a writer but in his 1921 passport application, he was a photographer. But perhaps the main reason for having only one story published that year was his forthcoming involvement in an exciting, entirely new adventure overseas.

---

[37] Ibid.

Chapter Six

White Waters

In 1921, Gordon MacCreagh was recruited to join the Mulford Biological Expedition which was intended to be a major, new scientific exploration of the Amazon Basin. It was organized by Dr Henry Hurd Rusby, a well- known American explorer, a professor at Columbia University and a staff member at the New York Botanical Garden. The venture was financed by the H.K. Mulford Company and it attracted great interest across the nation, with newspapers calling it the million-dollar expedition.[38] There were eight expedition members originally, including six eminent scientists - Rusby, Dr William Mann (an entomologist with the Department of Agriculture), Orland White (a botanist from the Brooklyn Botanical Garden), Dr Nathan Pearson (an ichthyologist) plus Frederick Hoffman (a statistician and ethnologist) plus George McCarty, a young taxidermist who was to also act as secretary and note-taker for the expedition. This collection of experts made the Mulford expedition the most highly qualified group ever to have set out on such an adventurous exploration. Finally, there was MacCreagh who was hired as quartermaster, photographer and general factotum plus a young American assistant.[39]

The expedition's ambitious mission was to explore the Amazon Valley from its headwaters in Bolivia down to the main part of the Amazon River in central Brazil. The original plan was to trek north-eastwards across Bolivia from La Paz, crossing the High Cordillera and then

---

[38] A pharmaceutical company based in Philadelphia that specialised in producing plant-based serums and antitoxins.
[39] According to the July 1921 report of the Philadelphia Academy of Natural Sciences, Dr A. Ruthven of the University of Michigan was also a member of the initial expeditionary group but his name does not appear in any other records.

descend using the local river systems to eventually reach the city of Manaus on the Amazon itself. This first part of the journey was expected to take some six to nine months to accomplish. The expedition would then rest and refit with new supplies shipped out from the USA, before continuing north-westwards along the Rio Negro and its tributaries, crossing back over the Andes to finally reach Bogota in Columbia. The second part of the journey was forecasted to last a similar time-frame. The expedition would travel through high mountains (over 18,000 feet), sparsely populated altiplano and an area of largely unexplored, dense jungle that was perhaps the size of France. We don't know how MacCreagh became involved with this prestigious venture; whether he pushed himself forward or he was invited to join. Certainly by 1921, MacCreagh was becoming increasingly known in the US as something of an expert on native people as well as a man with valuable experience of travelling through difficult jungle and mountainous terrain. His membership of the New York Adventurers Club would have brought him into contact with a wide range of people and he may well have heard about the proposed expedition by this route.

The Mulford expedition was by no means the first to try to explore the mysteries of the Amazon Basin. Over the years, there had been many previous expeditions, both exploratory and scientific, some successful and some not so. The first truly scientific expeditions were those led by the Germans, Alexander Von Humboldt in the early 19th century, followed a few years later by Von Spix and Von Martius who made huge botanical and zoological collections in the Brazilian Amazon between 1817 and 1820. The world's rarest bird, Spix's Macaw is named after Von Spix. Another notable scientific expedition was that of the intrepid English explorer and entomologist, Henry William Bates, who ended up spending an amazing eleven years in the interior of Amazonia between 1848 and 1859. He amassed the largest, single collection of insects ever made by one individual in the region, collecting almost 15,000 different species of which half were previously unknown to science.[40] During his first four years in the Amazon region, Bates collaborated with his close friend Alfred Wallace, co-discoverer of the theory of natural selection, although Charles Darwin published the theory first. When Wallace returned to Britain from the Amazon, his ship caught fire, and though he escaped with his life, sadly for him, all his collections and notes went up in smoke and then down into the depths. On his own return voyage to England in 1859, Bates made sure his collection was sent home in three different ships so as to avoid the unfortunate earlier fate of Wallace.

Almost a decade before the Mulford party set out, the former US President, Theodore Roosevelt had led a scientific expedition to the Amazon, together with Colonel Candido Rodon, a Brazilian explorer. Also sponsored by the American Museum of Natural History, they had surveyed parts of the lower Amazon basin and collected many new animal and insect specimens. However, the expedition was fraught with problems such as high fevers, constant sickness and a lack of food supplies. Roosevelt nearly died from an infected leg wound during the journey and was fortunate to return to the USA safely.

At the time of MacCreagh's first journey to South America, there was one other well-known British explorer of the Amazonian region, Lieutenant-Colonel Percy Fawcett. He was an artillery officer in the British Army and had first visited Brazil and Bolivia in 1906 to carry out mapping on behalf of the Royal Geographical Society. Fawcett returned to South America a further six times prior to the departure of the Mulford expedition and became increasingly fascinated with the rumours of a so-called lost city, lying somewhere deep in the Mato Grosso

---

[40] Bates' subsequent book about his travels, *A Naturalist on the River Amazons*, is a remarkable example of natural history writing and has been re-published several times.

jungle of Brazil. As the Mulford expedition members began assembling for their trip in 1921, Fawcett had just returned to England after an unsuccessful solo attempt to find this fabled lost city, having suffered from fever and forced to shoot his pack animal. The problems encountered by Roosevelt and Fawcett underlined the potential physical dangers facing MacCreagh and his fellow travellers and the importance of careful preparation and organization.

Fawcett was to return to the Amazon to pursue his search in 1925 accompanied by his eldest son, Jack and one of his son's close friends called Raleigh Rimmel. Although their expedition was well-planned and supplied, they disappeared into the uncharted depths of the Brazilian jungle and none of them ever returned. There has been much speculation over the years as to exactly what happened to the Fawcett group but according to most sources, they were probably killed by one of the fierce local tribes known to live in the area.

MacCreagh and his young assistant were sent on to La Paz via Peru a few weeks ahead of the main party to organize the purchase of supplies and mule transport. Meanwhile, when the ship carrying Rusby and his companions attempted to set sail from Brooklyn docks on June 1st 1921, waved on by a large crowd of family members and colleagues, it was discovered that the vessel had been seriously sabotaged the previous day by striking New York ship's engineers. They had even stolen some of the ship's machinery blue-print plans to handicap any attempted repairs. The ship, together with the frustrated expedition members, was forced to wait at anchor for a further day until marine engineers from New York arrived to sort out the problems. It was an ominous and embarrassing start for the much-hyped, million-dollar expedition.

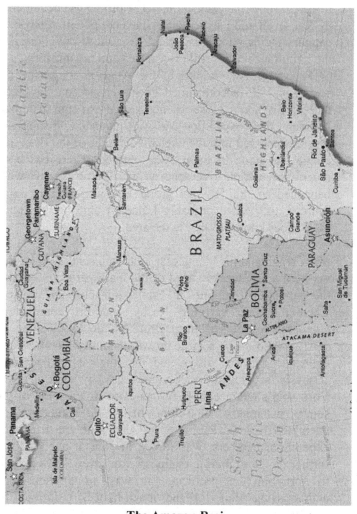

The Amazon Basin

After the initial false start, the remainder of the voyage southwards was relatively trouble-free and the full Mulford expedition assembled in La Paz, the capital of Bolivia in July 1921. They quickly moved up into the high Andes to begin searching for specimens. Once they reached the small settlement of Espia on the Bopi River, they were able to use balsa-wood rafts and slowly paddled down the river, stopping for weeks at a time to collect in the areas around Huacho and Rurrenabaque. Things went well for the initial few weeks but by late August, various problems had started to arise. First, while crossing a river, the party lost five boxes of vital supplies, including some of their provisions and most of their precious ammunition. The ammunition was important as the expedition was depending on it to obtain museum specimens of rare birds and small mammals as well as supplying the camp with fresh meat and protection when travelling through areas inhabited by hostile Indian tribes. Secondly, due to a variety of reasons, ill-tempered arguments broke out between the scientists and in particular, open dissent regarding Dr Rusby's leadership. Then, in December, Rusby, who was sixty-five, was forced to leave the expedition due to ill health, largely brought about by severe neuritis from an infected tooth. He was left behind to travel alone down the river until he was rescued by a government patrol boat. Before he left, he hired Martín Cárdenas, then a young botany student in Bolivia, to assist with collecting specimens.

William Mann then took over the expedition's leadership, while White, Pearson, Hoffman and Cárdenas continued collecting specimens in the lowlands of Eastern Bolivia. However, the scientists never reached the main part of the Amazon. After a series of mis-haps, they decided to head home from Cachuela Esperanza in Bolivia, in mid-March 1922, after only eight months in the field. Despite cutting short their expedition, they returned with over 2,400 items representing more than 1,500 species of plant life. Their impressive collections included a large number of orchids, termites, economically valuable plants and seeds. Back in New York, Rusby worked through most of the samples himself, identifying six new genera and 257 new species. The main collection from the expedition was eventually split between the New York Botanical Garden and at the Brooklyn Botanical Garden.

Although the expedition achieved many things, it was prematurely terminated and failed to reach its primary, overall objective. The harsh climatic and topographical conditions took their toll on the scientists who were physically and mentally unfit for this type of prolonged journey. The expedition members had been selected for their scientific standing, not their field experience or physical endurance. It also turned out that "In temperament they were, to put it mildly, unsuited for enforced close association and collaboration. It was a team united by distrust, insensitivity, and overblown egos." [41] From the beginning it was an expedition fraught with problems, some of which could not have been foreseen by the scientists but many were of their own making. It is perhaps significant that although various articles about the expedition were subsequently published and Rusby kept a diary which was typed up daily, there was never a formal, official report.[42]

During the expedition, MacCreagh kept detailed notes on what he saw and experienced and on his return to New York, he gradually worked these up into a personal account of the expedition which was first published in 1926 with the title of White Waters and Black.[43] Because

---

[41] George Schaller, eminent field biologist and anthropologist – foreword to the University of Chicago Press 1985 edition of White Waters and Black.
[42] Rusby gave his diary, entitled *Slowly Down the Mountain*, to the New York Botanical Garden.
[43] Originally published by The Century Co., in 1926, several editions and reprints followed in both the USA and abroad. Doubleday Anchor Books released a paperback edition in 1961 and in 1985, the University of Chicago Press brought out a facsimile edition of the original 1926 edition. It reprinted the book once again in June 2001 to excellent reviews.

he was extremely critical of the leaders of most of the expedition's other members, he was careful to avoid any mention of the expedition's actual title or the real names of the scientists. Instead, he used pseudonyms, referring to the disorderly scientists sarcastically by such titles as the Director, the Entomologist, the Botanist, the Statistician, the Ichthyologist and the Scribe. He evidently felt that otherwise, his descriptions of their activities and petty behaviour could have left him open to a series of libel suits. His book is really a fascinating description of his adventures in South America rather than a serious scientific publication. Indeed, MacCreagh explains in the very first chapter of his book that he will not "encumber it with a single item of scientific value". As such, the book forms a natural progression from MacCreagh's earlier fictional adventure stories. As many of these were based on his real-life experiences, the transition to non-fiction was probably not too difficult to achieve for MacCreagh. Certainly, the book is both well-written and entertaining, with a light touch and the author's usual occasional injection of irreverent and often, self-deprecating humour. MacCreagh's style of writing grows on you. At first, it can appear a little juvenile, both in vocabulary and structure. But gradually he pulls the reader ever deeper into his story, painting engrossing verbal pictures of what he sees, the expedition characters he is with and of course the many adventures and difficulties that he experiences.

In his book, MacCreagh's describes how his initial role was to arrive ahead of the main expeditionary group to purchase local supplies, including the mules, handlers and guides needed to cross the Andes and to make their way through the uncharted jungles. He was also to secure the appropriate government permits and letters of introduction. He duly organized all this, acquiring some two tons of supplies in readiness for the group's arrival. He was astounded to then learn by telegram that the scientists, gathering in New York, were bringing a further four tons of supplies and equipment, adding to the two tons he had already purchased. MacCreagh sarcastically mused that he had once "explored half of Asia with less than one hundred pounds of supplies." The problem was compounded once the extra four tons arrived in Bolivia when MacCreagh found that none of the 104 boxes had any indication of their contents. Days were wasted (and tempers frayed) in having to open each box and check its contents. When the bloated caravan finally got under way, it had one hundred and seventeen mules, and stretched in serpentine fashion for more than half a mile. Inevitable straggling as the mule-train headed into the mountains often doubled and tripled the caravan's length. It proved to be a bizarre and unusual journey.

MacCreagh was also tasked with planning the expedition's route from La Paz to Manaus but it soon became apparent that there were no adequate maps of the area that they were intending to head into, despite all the pre-planning by Rusby. A meeting with the Bolivian Minister of the Interior to review the government's own maps was of little help – the limited information available was clearly highly suspect. As MacCreagh related in his book, at one point in their discussions, the Minister scratched out a section of some two hundred thousand square miles on the map before them, exclaiming it was all "Inexplorada!", unknown territory. The Minister finally recommended that MacCreagh should create his own maps as the expedition went along and asked for details to be sent so his country's maps could be brought up to date. Although this lack of knowledge about the territory was unsettling for the scientists when they arrived in La Paz a few weeks later, it was of not a major concern for a seasoned adventurer like MacCreagh. He rationalised the situation in his book with the simple axiom "that where nobody has been before, somebody may find something that nobody else has."

Shortly before the other expedition members were due to arrive in La Paz, MacCreagh decided to undertake a short trip into the mountains to search for a possible pass, accompanied by his young American assistant. It turned out that despite his youth, MacCreagh's assistant found the high-altitude trekking in the mountains far too strenuous and on their return, he fled the country in a panic, never to be seen again. MacCreagh was forced to find a replacement and eventually recruited Walter Duval Brown who had been the American Consul in La Paz. It was hoped that his knowledge of Spanish and local experience would smooth the way for the expedition as it made its way through Bolivia and across the Andes towards the Amazon. MacCreagh also hired a local guide and a Chilean cook to accompany the expedition as far as Manaus.

As well as sorting out the expedition's route and four-legged transport arrangements, MacCreagh was also responsible for buying the guns they would take with them on their journey into the unknown. It is clear that this particular task was something that MacCreagh found much more interesting and his considerable previous experience with guns is shown by his detailed, animated description of his purchases for the expedition.

"Rifles: four, one to each two men. Savage 25-3000—pretty, pretty guns! Balanced so that one can shoot with one hand, which I maintain is the first requirement for unknown country where sudden things may happen. Trajectory, owing to the phenomenal velocity, is point-blank up to three hundred yards, which makes snap-shooting a snap. .... Yet muzzle-impact, owing to that same velocity, in spite of the light-caliber bullet, is nearly a ton. Sufficient to knock endways anything in all the Americas, no matter where one hits it. Shot-guns: four, one to each two men, to interchange with the rifles. Stevens, sixteen-gage. Hammerless, of course; for hammers in the jungles gather twenty pounds of trailing vines per minute. A good serviceable gun without any frills to it. I should have preferred twelve-gage; for sixteen is feeble on water-fowl and has poor range. Revolvers: Colt's army .38. Thirty of them. I carry a luger automatic pistol. It is good enough for snap-shots on deer and brush turkeys, and frequently obviates the necessity of carrying a rifle. A noble and an inspiring battery, of which I am proud."

However, as MacCreagh would later discover, none of the other members of the expedition had ever handled a gun. As the expedition prepared to set out, MacCreagh wastes no time in letting the readers of his book know what he thinks about the other expedition members:

"eight white men; six of whom are professors of eminent standing in their respective branches of science; five of whom have never seen a jungle nor known anything about travel other than in trains; two of whom are men well past middle age and set in their ways; one of whom is known to be the typical cantankerous professor of fiction and the stage; and another one of whom has a well-established reputation as a college disciplinarian. This last one is to be the leader or, as he prefers to call himself, the director of the expedition."

And then,

"Such an expedition, it seems to me, will contain all the elements necessary to startling drama. Some of the drama will be comedy; some of it, beyond any manner of doubt, will be tragedy. The doings of these eight white men with their various idiosyncrasies, herded together in the enforced close association of jungle travel, will be worthy of record. ... So surely, too, will it be an interesting record. At all events, a novel one. For nobody has ever been sacrilegious enough to keep a running record of the intimate doings of a party of eminent professors loose in the wild woods."

However, he is at this stage, full of respect for their bravery and determination in undertaking such a challenging journey:

"tenderfeet who unhesitatingly embark upon a journey into the unknown Amazon jungles for the love of their science are people who have a moral courage which almost gives them the right to a sympathetic consideration of their idiosyncrasies.... Little hardships which are to me but the cussedness of any long travel, must be sore trials indeed to some of these less experienced gentlemen.... deference is due to men who are willing to sacrifice so much for the furtherance of human knowledge and for no material remuneration other than the pittance that is paid to college professors."

Although extremely arduous, the initial trek over the high cordillera of the Andes passed without any significant problems. Altitude sickness affected some in the party as did the long, hard days of walking along boulder-strewn paths. However, MacCreagh was not bothered by the conditions; it was an adventure and an experience to be enjoyed. "I flatter myself already that I shall be less affected than some, at least, of the others, for two perfectly logical reasons: (1) because I have done much of this sort of travel before; and (2) because I like it and have come on this expedition with the firm conviction that I am going to have a good time." Only one mule was lost in an accident along the mountain tracks but it happened to be the one carrying MacCreagh's precious set of bagpipes that he had brought along for some musical entertainment during the journey. A long delay ensued while, at MacCreagh's understandable insistence, his bagpipes were recovered from the precipitous ravine into which the mule had fallen. The unfortunate dead mule was not retrieved.

According to MacCreagh's version of events, the group had not been on the trail very long before disagreement among the "scientificos" began. A combination of the hardships and rigours of the journey, exacerbated by their individual egos, caused petty but regular arguments, largely about the group's speed and direction of travel. As a consequence, the scientists were increasingly at each other's throats as the expedition progressed, sometimes not speaking to one another for days on end.

After a couple of weeks trekking with the mules, the expedition arrived at its first significant staging post, the small town of Espía which sat at the headwaters of the

Bopi River. As planned, the mule train now returned to La Paz, leaving the expedition to arrange onward transport down river on local native rafts. However, due to some poor decisions by the party's leader, Dr Rusby, the group found themselves stranded on a barren sandbank with their six tons of baggage for several weeks. Their situation was exacerbated by the fact that the muddy river water at this location was polluted and undrinkable. They passed the time by unpacking and re-packing all of the 104 unlabelled crates that the mule train had carried from La Paz. In doing so, it became apparent that some of the vital scientific equipment needed by the Statistician had not been brought and he flew into a rage, denouncing the capabilities of the Director. Then, as if to add insult to injury, MacCreagh discovered that although there was an inflatable canvas canoe plus an outboard engine in one of the crates, Rusby had forgotten to purchase any petrol for "the best-equipped expedition" in the world. Finally, it turned out that there were neither any cooking pots nor any kerosene lamps in the crates. Rusby's standing within the group descended to a new low as did the morale of the whole party.

Eventually, the marooned expedition was discovered by passing Indians who organised to transport them down river in a small fleet of native balsa-wood rafts to the next settlement of Huachi. It proved to be an exhilarating ten-day voyage down the rapids of the fast-flowing Bopi River. MacCreagh later described it as being "like a prolonged series of tremendous Coney Island shoot-the-chutes combined with the Dragon's Gorge." [44] He was full of admiration for the remarkable boat-handling skills of the fearless natives who safely delivered the expedition members and most of their gear through one of the most exciting journeys that could be imagined. Although the natives made light of the perils of this long and difficult river journey, the dangers from capsizing in the fast-

---

[44] The Dragon's Gorge in Eisenach, Germany, is a steep, narrow ravine some 3 kilometres long. Presumably, MacCreagh had visited while studying in Germany.

flowing water were very real. In the end, the only mishap was the loss of some of their provisions and most of their ammunition when one of the rafts partially disintegrated after being swept into some rocks.

The group lingered a while in Huachi in order to sort out their baggage after the river trip and to allow the scientists some time to explore the area. MacCreagh took the opportunity to pursue his favourite pastime – hunting with the local Indians for tapir, deer and jaguars. After several weeks in Huachi, the Statistician, Frederick Hoffman, decided to leave the expedition. His disagreements with Dr Rusby and more recently, the Entomologist, together with the many problems encountered on the journey so far, plus the absence of his equipment, persuaded him that he had endured enough of "this fiasco of an expedition". He had been told by the locals that they could transport him on one of their balsa rafts further down the Bopi until it joined the Beni River where he could then connect nearby with a steam-driven launch that would take him on to civilisation. He seized his opportunity to leave with joy according to MacCreagh and was now the second man to quit the expedition.

Shortly after Hoffman's departure, it became clear that Rusby had an ulcerated tooth. He had been suffering for several days but now the pain was extreme. First MacCreagh (who refused) and then the Entomologist were prevailed upon by Rusby to try to extract the troublesome tooth. Once again, the lack of foresight in terms of planning the expedition's supplies became evident – there was no painkiller or disinfectant in any of the three medical crates. This omission by the Director, a qualified medical doctor, resulted in him personally suffering for his own inadequate preparation. Fortunately, they did have a pair of forceps, left behind by the recently departed Hoffman. With these, the Entomologist attempted an extraction but after a minute or two of trying to jiggle out the tooth, the pain was too great for Rusby and he signalled a temporary halt to the process. "After a horrid period of recuperation" as described by MacCreagh, "the Director opened his mouth again and the struggle was renewed; to terminate this time by the tooth's breaking off short at the neck, leaving the frayed nerve terminals hanging in full view." For the next few weeks, Rusby tried to treat his exposed nerve with some iodine provided by the Botanist but it only provided limited and temporary relief.

This wasn't the only medical problem at that time. First, MacCreagh became ill with fever – he had refused to take the daily dose of quinine used by the rest of the group as he hated the taste. As an experienced traveller in the tropics, he recognized the early symptoms and with careful treatment, was back on his feet after three days. Then, McCarty, the Director's secretary was bitten by a tarantula as he washed himself which resulted in painful swellings all over his body. Apart from mosquitos and poisonous spiders, there were several other annoying pests that the party had to contend with on a daily (and nightly) basis in the jungle. Minute flies that can get through any ordinary mosquito-netting and which leave a little purple blood blister where they have fed. Tiny bees that hover annoyingly in front of one's face and dart into the eyes. Sand fleas that deposit eggs in tender portions of the foot where they develop in a little sac about as large as a pea and have to be cut out. Finally, a species of blow-fly that lays its eggs in moist, sweaty clothing. Its larva, known as the screw-worm, when hatched out by the heat of the body, immediately bores in under the skin. There, it develops into a grub one inch in length, and as it digs to enlarge its living-quarters, the sensation is exactly like that of being drilled into with an auger. As the expedition later penetrated further into the Amazonian Basin, yet more nasty creatures awaited them in the shape of hungry piranhas, stinging black hornets, voracious ants and dangerous caymans.[45]

The next leg in the expedition's journey was another lengthy raft trip down the Bopi River to the village of Rurrenabaque which Rusby had visited before on his previous trip to South America thirty years earlier. Once there, Rusby had arranged for the group to be met by his old agent, Miguel Bang, who would make all arrangements for their subsequent excursion into the interior. There were only a couple of hundred inhabitants of the village and very little of note but for MacCreagh, it proved to be an interesting place. At each stop along the way, MacCreagh always seemed to find time to talk to the local people, gathering information on their history, traditions and customs. Like all good travelogues, it is often the fascinating characters encountered along the way that bring added interest to the story and Rurrenabaque had more than its fair share. MacCreagh's abilities as a linguist are also evident as he gradually picks up enough words in the local Indian dialect to be able to hold a basic conversation.

During the meeting with the Bolivian Minister of the Interior in La Paz some months earlier, he had asked the expedition to check on the presence (or not) of a large lake marked on the government map as Lake Roguagua. Its exact location, size or even existence were uncertain but it was thought to lie near Rurrenabaque. The Minister's request was agreed and added to the expedition's itinerary. However, when the group arrived, they found that Rusby's ineffective agent, Bang, had made no arrangements for them to travel to the lake. A furious MacCreagh eventually found a local guide and hired several ox-wagons to take the expedition on the next part of their journey of discovery. After several days precariously riding the ox-wagons, they arrived at the lake and having explored and

---

[45] The cayman or caiman is a type of alligator common in the Amazon basin.

mapped it, found it to be of little scientific or economic interest or significance. Dissent and argument among the 'scientificos' erupted once again along the way, mostly involving Rusby who was in a foul mood, largely due to his tooth-ache. Rusby's health continued to deteriorate on the return journey and it was evident to all that he could no longer endure the rigours of the expedition. Although he bravely tried to carry on for a few days, "he was reduced to an agonized hobble, with bent back and swelling limbs." When the monsoon season then began, the combination of high humidity and excessive heat, meant that the game was up for Rusby. It was agreed that he should be immediately taken back to Rurrenabaque from where he could travel by steam launch to the British-run hospital in the town of Porto Velho for treatment.[46] Once sufficiently recovered, he would be able to take an ocean steamer to Manaus and eventually return home. After almost a year of travelling, he was the third man to leave the expedition.

With the departure of its director, the expedition's future lay in serious doubt. Before he left, Rusby had appointed William Mann, the Entomologist, as his replacement with instructions for the group to slowly make their way down river while continuing to collect specimens. But the remaining members were split as to whether to continue with the original goal of reaching Manaus and then travelling on to Columbia or to immediately return home. An additional complication arose when the Botanist, Orland White, adamantly refused to work under the direction of the Entomologist and so disagreement and dispute ruled their camp in

---

[46] The hospital was the remnant of a much larger complex built by the British during the glory days of Amazonian rubber. It sat at the end of a two hundred and sixty-mile long railroad track used to transport rubber latex down to the Amazon. At the time of its construction, it was said to be the most expensive railway in the world, costing some $37,000,000 and was now largely unused.

Rurrenabaque. In the end, the three remaining eminent scientists, Mann, White and Pearson, the Ichthyologist, all decided to quit the expedition and head home as soon as all their collections could be packed up. They were now the fourth, fifth and sixth men to leave.

Chapter Seven

Three Men in a Boat

When news came that the local steam launch had returned from delivering Rusby down-river, MacCreagh, together with Duval and young McCarty, decided to follow in the Director's footsteps to Porto Velho and from there, travel on to Manaus as originally planned. Although the scientific aspects of the expedition were now clearly prematurely terminated, for MacCreagh and McCarty, there was still a job of exploration to be completed and despite the hardships encountered so far, their spirit of adventure carried them forwards. As they made their way down to Manaus, some two months after Rusby's trip, the three men encountered an ever-increasing barrage of unexpected hostility from the people they met. It seems that as Rusby had passed through each village and town, he had left behind a trail of bad-tempered, vindictive complaints about "the frightful gang of crooks and spurious scientists" who had accompanied him on the expedition. At each stop along the river, MacCreagh and his two companions found themselves having to explain what had really happened in an almost apologetic, counter-narrative to that of Rusby.

As well as coping with the constant hostile reception as they continued downriver, MacCreagh now had a more serious concern. For several weeks, he had been suffering from an unknown tropical parasite that had burrowed its way into his leg. Despite treating it with quinine and trying to cut it out, the creature remained firmly lodged behind his shinbone, causing great discomfort and leaving his leg swollen and puffy. Perhaps mindful of the near-death of Teddy Roosevelt from a similar limb infection during his earlier Amazonian journey, MacCreagh was highly focussed on reaching Manaus as soon as possible in order to find medical treatment for his leg. By the time the

American trio reached the city, MacCreagh's leg was so swollen, he could only walk with the aid of a stick.

Fortunately, salvation for MacCreagh (but for not the parasite) was found at a British-run research station for tropical diseases. After two days of careful treatment by the medical team, the offending creature was finally removed from MacCreagh's leg. It turned out to be almost two inches in length with rows of tiny, black spines that allowed it to cling on tightly to its host's flesh. Much to the chagrin of the British doctor, the nasty parasite did not survive the removal operation but MacCreagh's leg gradually recovered and resumed its normal shape.

With MacCreagh now on the mend, the trio had to decide whether or not to follow the Mulford Expedition's original plan and head up the Rio Negro and then make for Bogota. As before in La Paz, consulting with the locals provided little substantial information about the possible journey ahead, other than it would be extremely difficult. The river along which they proposed to travel was full of rapids, there were tribes of unfriendly Indians along the way and they would again have to cross high mountains. Indeed, the American Consul's advice was not to try to make the journey at all due to the significant dangers. The area was full of bandits and savages he told the trio. After much debate, the three of them decided to push on, with MacCreagh as leader, in full recognition that whatever happened would be on their heads alone. When they arrived in Manaus, they found waiting for them the four tons of additional, new supplies that Rusby had ordered for the second leg of the expedition. Given that the little information they had gleaned so far about the journey ahead indicated it would be tough going, they decided to travel light and accordingly, MacCreagh arranged for the four tons of supplies to be shipped back to the USA. The only 'extras' that they would carry with them were a variety of trading goods to pay their way with the local Indian tribes such as knives, machetes, axe-heads and fish-hooks.

McCarty and Duval (centre and right)

MacCreagh Playing Bagpipes to the Tiquié

Following a few weeks of rest in Manaus, the trio set off on a steam-boat to travel up the mighty Rio Negro, some four miles wide at Manaus, with its banks covered in dense jungle. It was a six hundred-mile trip and took the steam-boat a week, battling against the current, until it reached its terminus at the small river port of Santa Isabel. Any further travel up-river would now have to be done by canoe and MacCreagh eventually managed to hire a large one, paddled by eight sturdy Indians. They spent a week heading upstream in the canoe until they reached São Gabriel from where the natives informed them, they would have to traverse some forty miles of rapids. Moving beyond São Gabriel initially proved problematic – there were no canoes to be hired locally. For several days, they explored different possibilities for moving up-river without any success. But then the river gods smiled on the party and almost simultaneously, an Indian arrived in a large canoe plus a small group of Columbian traders who were fortuitously also heading up-river. MacCreagh negotiated with the two parties and after some haggling, the Indian agreed to take the three men further up-river in his canoe while the Columbians would transport the American's equipment in their boats. Once the arrangements were all settled, it was agreed the small fleet would start out at dawn the very next day.

After a good night's sleep, MacCreagh and his two companions awoke to find that the Columbians had already departed, leaving a note to say that the Americans should follow as quickly as possible. When they went to find the canoe, they discovered that somehow in the night it had been damaged and now had a large hole in the bottom that would take days to repair. Annoyed that they could have been duped so easily, the trio were in a very awkward position. All their trade goods, personal equipment and cameras had disappeared up the Rio Negro with the crafty Columbians and there seemed to be no chance of catching them. However, the river gods were still smiling and having scoured the village, MacCreagh found an old, rusted-out motor launch whose engine, according to the owner, hadn't run for many years. The owner agreed that if MacCreagh could mend the boat, he could borrow it. The next twenty-four hours were like a frenetic scene from an Indian Jones movie. MacCreagh managed to borrow some tools and while McCarty and Duval rushed off in search of fuel, he immediately set to work to fix the engine. After several hours of concentrated effort, he eventually succeeded in getting the engine to run and the boat to float.

The good guys now set off in pursuit of the bad guys in true Indiana Jones style. However, finding the Columbians would not be easy; the massive river offered a hundred islands and a thousand creeks in which they could hide and evade the Americans. MacCreagh had been told there was one place where the river narrowed to pass through a deep gorge and if he was to catch the Columbians, that was the best place to do it. So, they motored directly there with as much speed as the decrepit launch could muster, to wait in the hope that the thieves had not already passed by in their canoes.

Although it was close to nightfall by the time the trio arrived at the narrows, the long journey had given the desperate but determined Americans time to formulate their plans and make their stand. They silently waited through the night with no sign of the Columbian boats. They waited again early the next morning in a thick mist that hung over the river. Soon after the sun broke through and the mist lifted, the Columbian canoes were seen paddling quickly towards them. They were astonished to find the Americans waiting ahead of them but MacCreagh and McCarty, armed with their rifles, soon forced the bandits into the shore. The Columbians initially expressed great surprise and indignation at being treated this way by their 'partners' and claimed to know nothing about the damage to the American's canoe. But they eventually accepted that they had been caught out, admitted their guilt and returned the Americans' possessions to them. For a brief moment, MacCreagh considered whether to impose some sort of retribution for what they had done but as "the poor devils had nothing" he simply let the Columbians continue on their way, glad that they had safely returned all their goods.

For the next few weeks, the mini expedition's progress up-river remained slow. At each stage, they had to negotiate some new form of onward river transport (the decrepit launch having been sent back to its owner). Also, at each settlement, they had to spend time discussing with local chiefs the best possible routes northwards as well as the potential dangers that lay ahead. Stories about the ferocity of some of the Indian tribes along the river grew more frequent, especially the Tiquié tribe who only recently had killed two intruding rubber traders with blowpipes. Having had lady luck on their side in São Gabriel, the American trio's fortunes now changed. As they made camp along the river bank late one afternoon, they were suddenly abandoned by their Indian paddlers and guides. After a few shouts and cries of dismay, "with one rush they piled into the boat, shoved frantically off, and bent to the paddles as if in the shore mud they had come upon the fresh tracks of the devil himself. In a second, they were a blur on the dark surface of the water. In another they were gone, and only the splash and clatter of their oars came back to us, going hard and in wild confusion."

As MacCreagh soon realized, the guides had indeed come across what to them was as bad as the devil himself. The footprints in the sand at the river's edge signalled they were now in the territory of the dreaded Tiquié Indians That was the last the American trio saw of those Indians or of the boat and the sudden desertion left them in an extremely awkward and hazardous situation. They had no transport and there was no prospect of any other boats venturing into that part of the river in the foreseeable future. Although the Americans had their weapons, they would be of little use in the event of a surprise attack by the feared Tiquié hidden in the trees with blowpipes. The trio agreed the only way forward was to try to make contact with the tribespeople and to seek their help.

That contact came unexpectedly quickly, as the Tiquié had been carefully monitoring the presence of these strangers in their land for some time. The apprehensive Americans soon found themselves being escorted through the jungle to the Tiquié's village where they were led into the tribe's large, communal thatched hut. Once inside, they discovered that they were surrounded by a group of some fifty naked warriors, all armed with spears "monstrous with the dim highlights glinting from their muscles." Initially fearing for their lives, the Americans soon realized that for the moment, the Tiquié were in fact more curious than life-threatening. Fortunately, MacCreagh's interest during the past few weeks in the local tribal language and etiquette of dealing with each other, proved of great benefit. He was able to open a dialogue and gradually build trust with the men in the hut, assuring them that the American trio were not like the greedy traders who occasionally invaded their territory. MacCreagh was eventually able to explain their predicament and assure the Tiquié that the Americans were simply stranded and if onward river transport could be provided, they would quickly be on their way.

Unfortunately, as MacCreagh now found out, the Tiquié only had very small dug-out canoes that they used for fishing. The largest would only carry a maximum of four people without any baggage so a small flotilla would be needed to move the Americans and their equipment up-river. The Tiquié made it clear that they were not interested in providing such a fleet of canoes but they would carry a message down-river to try to seek help for the Americans. MacCreagh had little choice but to agree to this plan and a messenger was duly dispatched southwards along the river. Days of anxious waiting gradually turned into a couple of weeks without any response to their cry for help. By then, it was clear that even if some floating miracle arrived to deliver them, the trio could no longer hope to work their way up-river and over the mountains to Columbia before the next rainy season broke. That part of the expedition's plan would definitely now have to be abandoned.

Potential salvation appeared some days later when MacCreagh discovered an old war-canoe a mile up-river. Although it had a hole in its hull and had sunk into the mud, it was thirty feet long and made of solid mahogany. MacCreagh reckoned the boat was salvageable and so bought it from the Tiquié chief for an iron spoon. He then contracted with the chief to supply three strong tribesmen "for as long as I should need them for a lump sum of three pieces of cloth, fifty fish-hooks, and as many iron nails as I should have left over out of my store after finishing my boat." After several days of hard work, the war-canoe was repaired, re-fashioned and re-floated, complete with a canopy for shelter.

As their enforced stay in the Tiquié camp had dragged on, Duval had become obsessed with getting out of the Amazon as soon as possible. He had now been with the expedition for eighteen months, much longer than had been estimated originally when the group started out full of confidence from La Paz. Apparently, his affairs at home needed his urgent attention and they were much more important than fooling about on a worthless and unknown river. By chance, as MacCreagh's 'private yacht' neared its completion, a lone rubber trader is his canoe appeared, heading down-river. For Duval, this was too good an opportunity to miss; he could at last escape the horrors of the jungle and begin the long journey home. After some discussion, the trader agreed to take Duval with him and as MacCreagh, alongside young McCarty, watched and stood waving their farewells, he slowly disappeared down the river. He was the seventh man to leave the expedition.

Constructing the Ark of Deliverance

At some stage after Duval had left, the Tiquié village chief approached MacCreagh one day, asking him if he could teach the old Indian how to play the bagpipes. MacCreagh includes a photograph in his book of the chief with two young teenage women who were apparently offered to him in exchange for the tuition. The author doesn't tell his readers whether or not he passed on his musical knowledge to the tribe but MacCreagh's pithy caption to the picture reads, "It has been spitefully remarked that a man who would like to learn how to play a bagpipe would have no morals anyway."

Still not done with adventure, the remaining duo decided to put their mahogany boat to good use by exploring further along the backwoods of the Tiquié River which flows along the equator for some three hundred miles. The village chief spread word up-river to his fellow tribesmen that the Americans were coming and he assured MacCreagh that the pair would be welcomed. The chief was as good as his word. As MacCreagh later wrote, their "progress up-river was almost a triumphal procession" and at each new village, they were greeted in a most open, friendly manner by the curious tribespeople.

From the descriptions in his book, MacCreagh clearly had an affinity for the Tiquié tribespeople and their simple way of life. He found them easy to talk to, intelligent and well-adjusted to their environment. As the days passed in one of the Tiquié's encampments, MacCreagh learnt that a special event would shortly take place. This was the amazing ceremony of drinking the caapi which had never been witnessed before by any white man. The caapi is a native cocktail that makes the Tiquié men brave enough to face the ordeal of fighting off the local evil spirit known as the Jurupary.[47] To be invited to watch the ceremony was truly a unique honour.

---

[47] Banisteriopsis caapi is a South American liana-type vine belonging to the Malpighiaceae plant family, usually boiled with leaves from other psychotropic plants to produce a hallucinogenic drink.

One of these is Gordon MacCreagh

MacCreagh in His Hour of Madness

MacCreagh sets the scene in his book for his readers. The whole tribe spends the first day dressing up for the ceremony. Both the men and women decorate themselves with ornaments, feathers and as much paint as they are able to muster. The men appear particularly striking, using a criss-cross design painted all over their bodies in scarlet from the juice of a local berry plus elaborate feather head-dresses and armbands. These were highly colourful, using the brilliant green, blue and vermilion feathers of the Amazonian macaw, toucan and humming-bird species. With the approach of nightfall, the ceremony begins in earnest. The women, who are not allowed to touch caapi, separate out from the men, who start to gather round in small groups. Sitting opposite each other they take up a repetitive chant, translated by MacCreagh as "Now we are about to drink the caapi. Now we shall be brave. We shall not fear the Jurupary when we see him."

The chanting continued for an hour or more until the caapi began to be served to the men in specially-carved gourds by cup-bearers who each wore particularly elaborate head-dresses with, according to MacCreagh "a thousand-dollars' worth of egret plumes. Unusually, for such an important and secret tribal ritual, the Indians were happy for the two Americans to join in. That they were so willing, is a testament to MacCreagh's remarkable ability to build both trust and empathy with tribespeople like these. He described the caapi as being "a thin, almost colorless liquid, flat-tasting and bitter, concocted from the leaves of a vine. . . it is not intoxicating. At least, not alcoholically so. But it imparts a distinct and almost immediate exhilaration. In cold retrospect I can't call it courage. But one acquires a certain feeling of - well, devil-may-care irresponsibility."

Although not present at the ceremony, a more detailed description of the effects of caapi was later provided by Dr Rusby, the departed leader of the expedition.

"Within a very short time after drinking the decoction, there is a powerful effect on the nervous system and on the circulation. The cutaneous circulation is checked, as manifested by a strong pallor. The subject is restless, and occupies a standing position. There is an intensely anxious or fixed expression to the countenance, and there are convulsive tremors. This condition lasts but a few minutes and is followed by a violent reaction, in which the blood rushes to the surface and the man becomes highly or even violently active. Fear, and even prudence, is entirely destroyed and he becomes extremely active muscularly. He is ready to fight anything and anybody, or any number of enemies and suffering is disregarded. He rushes about and seeks an enemy with the utmost eagerness. This condition lasts for hours, and is followed by more or less exhaustion and somnolence. The nature of this somnolence is in doubt. Some accounts indicate it is a narcotic effect and that there are unnatural dreams and visions. Others indicate that it is the natural reaction following weariness or exhaustion. Mr. MacCreagh records that he had no desire to fight, and no unusual mental effects beyond that of a powerful stimulation and a desire to dance and otherwise engage in bodily and mental activity." [48]

As the evening wore on, both McCarty and MacCreagh drank liberally from the caapi gourds. Buoyed by their mutual sense of adventure and the effects of the caapi, both men soon found themselves sufficiently stimulated and without any inhibitions such that they decided to fully join in the ceremony. They stripped off all their clothes and after their naked bodies had been decorated with paint and a suit of feathers by some of the women, the two white men went native and began dancing and chanting with the Indians in a large circle. After a while, on a given whistled signal, the circle broke up into what MacCreagh described as "a snake-dance. . . in endless time with its rhythm the dancers stamp and weave and swing and coil unceasingly through the night, through the next day, the next night, the

---

[48] H Rusby article in The Journal of the American Pharmaceutical Association, Vol 13, 1924.

day again, and so to the third night - the dread night of their meeting with their devil face to face."

Although MacCreagh and McCarty dropped out of the energy-sapping dancing after the first night, they watched mesmerised, as the over-stimulated Indians continued with their interminable dancing. The vibrating noise and rhythm were constant and unceasing, ruling their spirits and excited nerves for three full days, despite their empty stomachs. Eventually, on the third night, total weariness overcame the effects of the caapi and the exhausted tribesmen began to steadily fall away, stumbling off to their huts. As the firelight dimmed, the two Americans eventually found themselves alone in the dark, still enervated and tingling from the caapi and the rhythm of the dancing that still pulsed through their bodies. For MacCreagh, the ceremony had been fascinating in its overall theatrical effect but also rather creepy. As he stood alongside McCarty in the silence of the dark night, he wondered whether the Jurupary devil had indeed been vanquished or was simply waiting in the shadows of the trees, ready to pounce on them.

Somewhat unnerved, the two men quickly decided to make a hasty retreat and regain the shelter of their boat on the river. As MacCreagh wrote, they had "been privileged to see what no other white man has seen." It was a quite remarkable achievement and experience. But MacCreagh's Amazonian discoveries were not yet quite over. Villagers' tales about a nearby tribe called the Cihuma people who followed a trail by scent like animals had aroused MacCreagh's interest. A few days after the caapi ceremony, he and McCarty therefore set off in a canoe with a couple of Tiquié guides to see if they could find these strange people who hunted by scent. After two days of paddling, they eventually came across a Cihuma camp alongside the river. MacCreagh's description of these people in his book is not terribly flattering "nomad semi-apes making a crude fish-trap. A squat, misshapen gang they were, with large bellies and thin limbs and low gorilloid foreheads and prognathous jaws. Dirty they were, too, with matted hair and greasy bodies, dirty by preference like monkeys."

Communication was difficult as the Cihuma didn't speak the Tiquié language but the American duo decided to stay the night with the Cihuma in their camp in order to test their alleged scent-tracking abilities the next day. MacCreagh and McCarty were invited to join in an evening meal that the tribesmen had prepared by stewing the carcass of a large bird. As it happened, McCarty was suffering from one of his occasional bouts of malaria and so went to bed early without eating. MacCreagh however, joined the Cihumas eating the meal that had been brewing in an old pot over the camp fire. In the morning, McCarty's fever had passed but MacCreagh felt extremely ill, vomiting frequently and sweating profusely. He had a burning pain in his stomach and he was too weak to move. It was obvious he urgently needed medical treatment.

The resourceful, young McCarty now took charge. He bundled MacCreagh into the canoe and together with their two Tiquié guides, vigorously paddled back down the river to the Tiquié's main camp. They covered the distance in some nine hours compared with the two days it had originally taken. With the help of a medical kit that was stored in their mahogany boat, MacCreagh stabilised himself but was still a very sick man and it was agreed that he should get to the hospital in Manaus as soon as possible. Unfortunately, the Amazon River capital was at least six weeks of normal travelling down-river. McCarty again stepped up to the mark and managed to hire a double crew of eight Tiquié Indians who paddled the Americans' boat in relays from early dawn to late into the night. It took five weeks on the river to reach Santa Isabel from where the Americans caught the steamer ferry to Manaus a couple of days later. For the whole trip, MacCreagh was unable to eat anything and survived on half a dozen eight-year-old, rusty cans of condensed milk that he eked out a spoonful at a time. This story is reminiscent of the experiences of MacCreagh's former flatmate, Captain Dingle, while sailing to Bermuda.

During their short stop-over in Santa Isabel, MacCreagh and McCarty gave away or traded most of their remaining equipment and guns, acquiring a variety of souvenirs and mementos to take back to the USA. Several of the items MacCreagh brought back were subsequently donated to museums and perhaps the most famous was a highly-decorated, Indian log drum or trocán. When MacCreagh had first come across this drum months before on his way up-river, the tribal chief who owned it offered to sell it for a thousand dollars. With MacCreagh evidently about to depart for good, the chief eventually agreed, after much haggling, to exchange the drum for one of MacCreagh's rifles. Despite his weakened state, MacCreagh recognized that he had been fortunate to have "got a prize indeed."

Once they reached Manaus and MacCreagh was safely under treatment in hospital, McCarty took his leave and returned to the USA. Of the Mulford expedition's original group of nine that had set out almost two years earlier across the Andes to explore the Amazon, MacCreagh was now the last man standing, or in his case at the time, lying down. It was a remarkable achievement and something of which MacCreagh was rightly proud. He was treated in Manaus by a British doctor who was a specialist in tropical illnesses who confirmed that he had indeed been poisoned by the Cihuma. MacCreagh remained in hospital on a strict diet of curdled milk supplemented by medicine prescribed by the doctor until, after some six weeks, he was well enough to make the return trip to New York. He finally left South America, accompanied by all the various mementos he had collected along the way, plus the "remnants and inhumanly voluminous reports of the expedition."

In addition to his many other roles during the expedition, we should remember that MacCreagh was also the official photographer. He took countless black and white photographs (many of which feature in his book) and filmed the expedition. As well as a still camera, he took with him a movie camera especially designed for the expedition by Carl Akeley,[49] a well-known African explorer and naturalist. Sadly, most of MacCreagh's reels of film, which the expedition's sponsors had planned to make into a motion picture, were lost when one of the expedition's boats overturned while navigating river rapids. Although a few reels did survive the expedition and were subsequently shown at museums and scientific conferences, they have since disappeared.[50] However, their existence was confirmed in various newspaper articles at the time.

When MacCreagh landed back in New York on August 19th 1922, he found himself at the centre of considerable media interest. He was feted as a hero and his remarkable achievements were quite rightly celebrated. A lengthy New York Times article published the day after MacCreagh's return proudly announced his safe arrival back in the city. It was headlined "Devil-Devil Dance Caught By Camera / Gordon MacCreagh, Back From the Amazon, Describes Torture Ceremony." The newspaper went on to describe how:

"Returning yesterday on the steamship Polycarp from South America, where for eighteen months he has been a member of the Mulford Biological Exploration Expedition which explored the Amazon Basin, Gordon MacCreagh, ethnologist, brought back with him a detailed description ... of the Caapi, a Devil-Devil dance which scientists have

---

[49] Akeley was a pioneering American taxidermist, biologist, conservationist and nature photographer. He died from dysentery during an expedition to the Congo in 1926.

[50] According to Peter Ruber's account of MacCreagh's life, the surviving films he took during the expedition were the object of an unsuccessful, decade-long search by Susan Fraser, then Head of Information Services and the Archivist of the New York Botanical Garden.

thought extinct, but which still flourishes among Indians in sections visited by Mr. MacCreagh."

"While three young wild cats played about and a macaw, gorgeously bedecked in many hues, shifted lazily from one foot to the other, Mr. MacCreagh, in his studio at 21 East Fourteenth Street, related the story of the Caapi which he witnessed on the Tiqui River among Indians which were conquered by the Incas on the Waupea, a tributary of the Negro, in Brazil. Mr. MacCreagh brought back with him some of the caapi, which he sent to Dr. Rusby for analysis. He also brought back other data and material which will take him some time to prepare in the way of a report."

The wildcat kittens and macaw referred to in the above article were part of a varied collection of creatures (alive and dead) plus various artefacts that MacCreagh brought back with him from the Amazon. The Buffalo Evening News of August 23rd, 1922 featured a photograph of MacCreagh in his New York apartment feeding the kittens with milk. He later gave them to the zoo in Central Park but apparently kept the macaw for some years.

A second New York Times half-page feature article followed on September 3, 1922. Luridly entitled "Devil Dancers", the newspaper included a fascinating, feature photograph of MacCreagh in the full costume of the Devil Dancers. Holding a bow in one hand and a ceremonial spear in the other, he is barely recognizable. He admitted in the article that during the several days the Devil-Devil Dance lasted, he had consumed a fair amount of caapi. Apparently, it had no alcoholic properties or a narcotic taste like opium, but it did give him a feeling of high exhilaration. This made him lose all self-consciousness to the point of allowing the natives to paint and dress him as one of the dancers – the details of which he did not remember too well. The article also described how poison-tipped arrows and blowguns were used to sacrifice animals during the dance ritual. "In spite of the terrible effects of the ordeal on them," the New York Times continued, "Mr. MacCreagh managed to get the

Indians to go through [the dance] in daylight for the benefit of the moving picture camera." Presumably, young McCarty was behind the camera as MacCreagh was shown participating in the ritual event in the film footage that was later brought back to the USA.

Then in a May 16, 1923 New York Times article headlined "Indian Picture Shown," the newspaper reported on the first public showing of his surviving movie films:

"For the benefit of persons interested in the Amazon Exploration Expedition, led by Gordon MacCreagh, ethnologist, films dealing with the Tiqui-Tucana Indians, a new tribe found on the Tiqui River, were shown yesterday in the American Museum of Natural History." The films were also subsequently shown at the December, 1923 meeting of the American Pharmaceutical Association but there is no record of any subsequent event. Several months after his return to America, MacCreagh donated a collection of various artefacts to the Museum of the American Indian in New York. The gift included items illustrating the ethnology of the Tiquié and Tucano Indians along the Brazil-Columbia border as well as the aforementioned trocán drum.[51] He also gave a collection of vertebrates and insects gathered in the Tiquié River region to the US National Museum in Washington DC in 1923.

Despite all the trials and tribulations that MacCreagh and McCarty shared during their time in the Amazon, it seems they lost touch with each other on their return to the US. However, MacCreagh did keep in contact with Dr William Mann, the expedition's entomologist, who had gradually gained MacCreagh's respect during the journey. The pair occasionally exchanged information on some of the more exotic insects and other creatures that MacCreagh found on his later travels around the world. Mann subsequently joined the Smithsonian Institute and

---

[51] Established in 1916, the museum eventually became part of the Smithsonian Institute.

then the National Zoo in Washington D.C. but didn't participate in any further overseas expeditions.

It is interesting to note that when MacCreagh's book appeared in 1926, it was the only record of the expedition that was published. Although the New York Botanical Garden was given copies of the manuscripts produced by Henry Rusby, the expedition's director, as well as records written by other scientific members, it has never seen fit to formally publish them. MacCreagh's book received extremely favourable reviews and became increasingly popular. The Washington Post Book World described it as "The highly unofficial account of an Amazon Expedition that might have been staffed by the Marx Brothers." On November 12, 1926, the New York Times wrote that the book was, "A running record of the intimate doings of a party of eminent professors loose in the wildwoods." The Explorers Journal called it simply a "classic in explorational literature." White Waters and Black continues to be widely read around the world and is often referred to and quoted almost one hundred years after its publication.

The considerable media interest shown in both MacCreagh's South American journey as well as his book, meant that he now found himself something of a minor celebrity in New York. Although this didn't go to his head and his feet remained firmly planted on the ground, his new status brought several subtle changes to his life. MacCreagh's standing and reputation within The Adventurers Club would certainly have risen, opening doors to new contacts and opportunities. His capabilities and self-taught knowledge gained in the field relating to wild-life brought him a certain level of respect within the scientific community. While he was no scientist, or a formally-trained professor like Indian Jones, his input and views were increasingly sought by others on a variety of matters in the general world of science. Lastly, perhaps for the first time in his life, MacCreagh was now in a reasonably sound situation financially, allowing him to freely focus on both his writing and further travel.

Chapter Eight

Ethiopia Once

Whether MacCreagh was a native-born or naturalised American, there is no doubting that by the early 1920s, his home was New York and his allegiance was to the USA. From now on, in his writing, he is clearly an American citizen and increasingly, his references to events and the heroes in his stories are American. Interestingly however, his adventure stories remain almost exclusively set in the familiar lands of Africa, Central and South America or the Far East. Although he travelled extensively in the USA over the years, the country and its history never seemed to provide him with any great inspiration for a tale. Perhaps he felt there were enough writers specialising in pulp-fiction stories about the Wild West.

Soon after MacCreagh returned to New York from the Amazon in August 1922, he met Helen Komlosy and within a year, they were married. Born in November 1894 in Boston, Massachusetts, Helen's parents had emigrated from Hungary. The family later moved to New York where her father worked as a scenery artist for the Metropolitan Opera House. Helen also started work in the city at the age of fifteen. When they first met, Helen was living on the East Side, not far from MacCreagh's apartment and she was employed as a milliner. She is described as being 5 feet 5 inches tall with a high forehead, blue eyes, brown hair, fair complexion and an oval face. By all accounts, she was a very striking and attractive woman with a lively personality. In April 1922, a few months before she met MacCreagh, Helen had travelled on holiday to France, Germany, Austria, Italy and Switzerland. She left New York on the *SS St. Paul* on June 7, 1922 and returned from Southampton to New York aboard the *Olympic* which arrived in New York on July 12, 1922. The fact that Helen had embarked on such a long, overseas journey is evidence of her interest in travel, something that she would now share regularly with her new husband. She also turned out to be an expert shot with a rifle, not something one would expect from a New York lady milliner. For both of them, marriage had come relatively late in life. She was twenty-eight and he was now almost thirty-four.

The newly-weds enjoyed life in New York for a couple of years and MacCreagh continued with his writing, though the flow of stories slowed down – as a newly-wed, this is perhaps understandable. However, he did write two stories for *Adventure* magazine that were clearly based on his Amazon experiences - *The Inca's Ransom* in 1924 and *The Creek of the Poisonous Mist* in 1926. Both were subsequently published by Chelsea House of New York as separate hardback novels.

These two tales are especially interesting because of their possible links to subsequent Indiana Jones movie plot-lines. *The Inca's Ransom* is set in the high Andes and relates the struggles of an ancient tribe of Indians in defending their civilization and treasure against a group of greedy White adventurers. In *Poisonous Mist*, MacCreagh takes his readers to a mysterious, hidden land, deep in the Amazon jungle which is the home of a long-lost civilization with supernatural powers. The close similarities between MacCreagh's stories and the later cinematic events seem too striking to be purely coincidental.

MacCreagh's urge to travel and go-look-see was still strong and it would seem that these two stories generated sufficient funds for him to contemplate a new adventure. The MacCreaghs decided to leave New York on a trip to re-visit some of MacCreagh's old haunts, sailing first to Britain and then on to Borneo. We have no information on how long they were away or exactly what they did and MacCreagh doesn't mention whether he brought back any snakes or orangutans this time. However, the experience of travelling in remote regions with her new husband was evidently not sufficiently daunting to discourage Helen MacCreagh from further such adventures.

In early 1927, the MacCreaghs set off on yet another exciting journey, this time an extended trip to Ethiopia to search for the lost Ark of the Covenant. MacCreagh never fully explained the reasons behind choosing to go to Ethiopia. Although he hadn't visited the country while he was in British East Africa years before, he would have been aware of its primitive remoteness and the abundance of wild life for good hunting. It seems he was also attracted by its ancient history and unique position as the only African country not to have been annexed or colonised by a European country. As well as the romance of exploring a little-known country and conducting a search for the Ark of the Covenant, the trip's objectives also included capturing specimens of local fauna (in particular, lions) and tracking down the Falashas, a legendary lost African tribe of Jewish descent.[52] The potential for some big-game hunting on the occasional side-trip during his travels was definitely an added attraction for MacCreagh. The expedition and its objective of finding the Ark caught the attention of the press, both regionally and nationally. A *New York Times* article dated January 1, 1927, quoted from an interview with MacCreagh:

"It is intended," Mr. MacCreagh said, "to climb Mount Nebo, where, according to legend, the ark is located. The story is that Menelik, the son of Solomon and Sheba, paid a visit to his father, who gave a reproduction of the ark to him. The boy is said to have tricked Solomon, stealing the real ark and carrying it to Abyssinia. One of the tribes to be visited [the Falashas] is descended from the Jews who quit the leadership of Moses at the beginning of the exodus, being sceptical of his power to part the waters of the Red Sea."

---

[52] The Falashas claimed to be descended from the Queen of Sheba and King Solomon and retained their Jewish faith when Ethiopia converted to Christianity in the 4th century.

Mount Nebo is mentioned in the Bible as the place where Moses first gained a view of the Promised Land of Canaan and the prophet Jeremiah is supposed to have hid the tabernacle and the Ark of the Covenant in a cave there. However, MacCreagh seems to be a little confused here as Mount Nebo is actually located in Jordan – a long way from the planned destination of Ethiopia. The search for the legendary Ark of the Covenant, one of the Bible's most prized artefacts, has been going on for centuries and continues to this day. The Ark is supposed to have been made some 3,000 years ago by the Israelites to house the stone tablets on which the Ten Commandments were written. It is described in the Bible as an ornate, gilded case, about the size of a 19th-century seaman's chest, made of gold-plated wood, and topped with two large, golden angels. It was carried using poles inserted through rings on its sides (see illustration on back cover).

Some archaeologists have searched for the Ark of the Covenant simply to prove the biblical narrative is true. A few have simply been good, old-fashioned treasure hunters since the Ark, if found, would certainly be a priceless artefact. Others have pursued the search because the Ark is linked to several of the Old Testament's miracles and is therefore alleged to have special powers. This was very much the scenario in the movie *Indiana Jones and the Raiders of the Lost Ark*, where the Nazis were looking for the Ark (in Egypt) to use it as a supernatural weapon. Although confused about Mount Nebo, MacCreagh was right however, to direct his search to Ethiopia as one of the strongest claims about the Ark's possible location was that, as he subsequently described, it had been brought to the country early in the first century AD. If MacCreagh was truly able to find the Ark's location, it would be a major feather in his cap as well as raising his international standing as an explorer.

In a subsequent article, *The New York Times* reported on the details of a slightly modified but still ambitious itinerary that the MacCreagh's had planned:

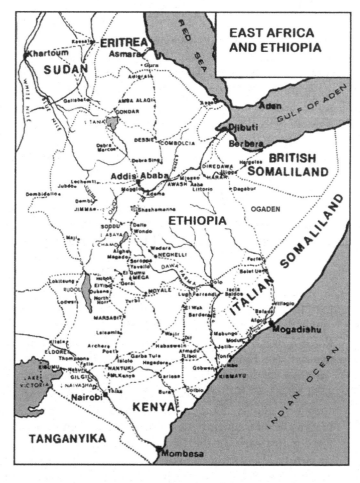

East Africa and Ethiopia

"Mr. MacCreagh and his wife, who accompanied him, plan to make their way overland across Abyssinia, over the Abyssinian Mountains to the Sudan plains and out to the Nile. From the capital of Abyssinia, Addis Ababa, the expedition will go north for big game hunting and the collection of specimens of wild life, then to the mountains near the source of the Takisazzee River, in the search for the Ark of the Covenant about which many native legends are told; and from there to the Sudanese border." [53]

MacCreagh must have been planning this trip for some time before they left. He not only had to sort out the extremely complicated arrangements for making such a difficult journey but he also organized funding from *Adventure* magazine, on the basis of providing a series of articles for the journal during his trip. In addition, MacCreagh arranged further sponsorship by signing up as a roving reporter for the *Saturday Evening Post*, one of the most widely circulated and influential US weekly magazines of the period. This not only helped finance the expedition but also, as an accredited reporter, it would give him added prestige. However, as we shall see later, members of the press were not especially highly regarded then in Ethiopia. MacCreagh subsequently remained a contributor to the *Saturday Evening Post* for many years.

He also negotiated with a Los Angeles film studio for a professional photographer to accompany the expedition to provide a film record of the trip. The man selected for this role was the extremely experienced Earl Rossman who was an explorer in his own right and had worked as a camera-man in both southern Africa and the Artic.[54] MacCreagh's aim was almost certainly to sell the movie footage to British Pathé, the world's leading newsreel and documentary producer of the time and a company that Rossman had worked for previously. Helen MacCreagh

---

[53] MacCreagh was probably referring to the Takkaze River.
[54] Rossman wrote and directed two films about the indigenous people of the Artic – Kivalina of the Ice Lands in 1925 and Dangers of the Artic in 1932.

evidently decided early on in the planning process that she would also accompany her husband on this trip; she was not one to be left at home to read magazines and play golf all day.[55]

The couple left New York on February 10th, 1927 on the *SS American Merchant* and after a rather tortuous journey via London and Paris (to acquire supplies and permits for the expedition), they sailed through the Suez Canal and some three months after setting out from the USA, arrived in French Djibouti. After a frustrating but predictable delay of several days at the port, they eventually caught the train for the leisurely, three-day overland ride to Addis Ababa. Part of the reason for the delay in Djibouti was the fact that the MacCreaghs had shipped with them twenty-two packing cases. At first sight, this seems an incredibly large amount of baggage for just two people but as MacCreagh later explained in his subsequent book about the expedition, *The Last of Free Africa*, they had brought with them everything needed for a sojourn of many months, travelling in relatively inhospitable country. MacCreagh knew full well that there would be little opportunity to replenish any of their supplies during the trip.

As always with MacCreagh, guns and ammunition form a vital part of his equipment, as do tents, saddles, medical kit, tools, clothes, food and cooking equipment. But as he ruefully comments, "anyone who has travelled with a lady can explain away any quantity of trunks." His wife's requirements evidently added significantly to the overall load. Both the guns and his wife's essentials proved to be the cause of various problems during the trip. Despite having the appropriate permits to import the guns and ammunition into Ethiopia, the local customs officials in Djibouti were suspicious of his intentions. Only by making several additional 'special payments' did

---

[55] Several US press articles of the time reported that Mrs Rossman also travelled with her husband to Ethiopia.

MacCreagh manage to continue on with his arsenal. One of the most important of Helen's essentials, was a copious supply of expensive cosmetics, carefully sourced in New York to provide maximum protection against the worst of the African sun.

Also included in their baggage was a set of long and short- wave radio equipment. The receiver was made in New York by Grebe & Co. while the short-wave transmitting apparatus was of British make and purchased in London, along with the recruitment of an experienced radio operator. On February 20, 1927, the *New York Times*, in another, longer article about the MacCreagh expedition proclaimed, "Explorer In Abyssinia To Use Short Wave Lengths / Gordon MacCreagh Expected to Communicate with American Amateurs From African Wilderness - Receiver Similar to Byrd's Polar Set." [56] In the article, the newspaper explained in some detail how:

"Amateur radio operators who have picked up messages from Arctic explorers and from the Dyott expedition[57] now following Theodore Roosevelt's trail down the River of Doubt through the jungles of Brazil are likely to hear dispatches from Gordon MacCreagh, exploring in Abyssinia ... The short-wave amateur radio station 2ZV, at Richmond Hills, N. Y., will attempt to maintain constant communication with the expedition...Engineers have expressed the opinion that the expedition should not have any difficulty in hearing American and European broadcasting stations."

"The apparatus will be carried across the Abyssinian Mountains on the backs of camels. Dry batteries will be used for supplying energy to the receiving sets; and a hand generator will furnish the high voltage for the short-wave transmitting set. The expedition will be in the wilderness

---

[56] Commander Richard Byrd, the famous American aviator and polar explorer who attempted to fly over the North Pole in 1926 and successfully flew over the South Pole in 1928.
[57] George Dyott was a well-known English explorer and pioneering aviator of the 1920s.

for about six months...It is expected that the transmissions will begin in about a month."

Despite all the press hype about this impressive new technology, in the articles and book that he subsequently wrote about his trip to Ethiopia, MacCreagh fails to mention anything about the radio equipment, camel transport or the operator hired in London. It's possible that the radio set was lost or damaged in transit to Ethiopia or simply failed to function on arrival. I have been unable to discover records of any transmissions being made during MacCreagh's trip and the fate of the radio operator is unknown. Similarly, I have not found any evidence of a film record of the expedition having been produced. Although MacCreagh does mention his cameraman, Rossman, in his book, his comments are very terse and it is clear that the relationship between the two men became increasingly strained. Given Rossman's previous considerable experience, it is hard to understand why. Since MacCreagh was also a capable photographer, there may have been some professional rivalry or disagreements about what and where to film.[58]

Although it seems that nothing came of the radio or film projects, the series of stories that he had agreed to produce for *Adventure* magazine were published. There were seven separate articles printed between June 1927 and May 1928 under the title of *The Abyssinian Expedition*. MacCreagh must have written these during his travels and then sent each one over to New York for publication. For those readers who pay attention to such things, it's worth commenting on the title of the articles that MacCreagh used. At the time of his visit, the country was known both as Abyssinia and Ethiopia but there are subtle differences in the origins and meanings of each name. The term Abyssinia is apparently derived from a rather derogatory Arab word meaning 'a nondescript mixture' which probably arose due to the variety of tribes

---

[58] According to MacCreagh, of the twelve thousand feet of film shot by Rossman during the expedition, just two hundred feet were eventually used by his company.

and peoples that have settled in this land over the millennia. When MacCreagh arrived, the country had long been ruled by a Christian, Amharic-speaking people who, perhaps understandably, disliked the contemptuous meaning of Abyssinia and so called themselves Ethiopians. In his writings, MacCreagh fluctuates between the two names, even though he was well aware of this distinction between them. For clarity and consistency in this book, I have used Ethiopia when writing about the country as well as when referring to its people, except where I have quoted directly from other sources.

Soon after his return to the USA, MacCreagh began to develop and expand the series of stories that he had written for *Adventure* magazine into his new book, *The Last of Free Africa* which was first published by The Century Co. in September 1928. The book is subtitled "*The account of an expedition into Abyssinia with observations on the manners, customs and traditions of the Ethiopians with some pungent remarks on the anomalous political situation that, at present, obtains between this ancient kingdom and the nations of the world.*" The book received generally excellent reviews on its publication and proved to be exceptionally popular with the public at large. The interest in MacCreagh's new book was partly due to the subject matter – travel books about Ethiopia were few and far between – and a fascination with the main purposes of the trip, locating the Ark of the Covenant and the Falashas. Its success resulted in the original edition being followed by several subsequent printings as well as a cheap mass-market version by Grosset & Dunlap. In 1930, Century also published a special edition of the book as part of its twelve-volume Vagabond Series celebrating well-known travellers. A British edition and several foreign translations then followed with a very popular, second US edition in 1935 by Appleton-Century Co. in which MacCreagh added a foreword and epilogue.

The review published by the Stanford Daily of Florida on February 18th, 1929 is worth quoting, not so much for its comments about the book itself but for the additional information on MacCreagh's past life:

"By this time the world must be as familiar to Gordon MacCreagh as his own back yard is to a Peninsula commuter. He has hauled teak out of the Himalayan forests, caught animals for Hagenbeck's zoo, served with the Burma opium police on the Siam border, explored Tibet and camped out for eighteen months in South American jungles. Abyssinia was about the only spot remaining, so into the heart of the ancient kingdom of Ethiopia he trekked — with a wife, and twenty-odd packing cases and a gay spirit of adventure. And it is with a blithe-hearted pen that he records his experiences. Mr. MacCreagh is more intent upon making a contribution to the reader's entertainment than to his scientific knowledge. Even so, he imparts a good fund of first-hand information in the course of these chapters and brings that distant region of the earth's surface vividly before one's eye. And besides, his camera tells much."

This is the first and only recorded mention of MacCreagh having worked either as a lumberjack in the Himalayas or with the Burmese opium police. As with so many aspects of MacCreagh's life, it is possible that there is some element of truth in these statements but we don't really know. The relatively short period that he spent in both of these regions suggests there was hardly sufficient time for MacCreagh to have done these things on top of everything else. However, I leave it to the reader to decide whether this is another example of the author's periodic, imaginative hyperbole when talking to the press.

In the opening remarks of *The Last of Free Africa*, MacCreagh sets out a slightly different agenda for his book from what one might have anticipated from the initial press coverage of the expedition. The book is about "The Unconquered. Free since the beginning of history. Governed by its own hereditary ruler ... descendant of King Solomon." The question of how Ethiopia "has remained, as has no other country in the world, free since all known time" clearly intrigued MacCreagh during his time in Ethiopia and now seemed to be the author's primary focus in writing his book. The original objectives of travelling through the mountains and on to Sudan, contacting the Falashas and searching for the lost Ark of the Covenant, appear to have become subsidiary activities during his travels. Much of the book follows the format and content of the seven articles MacCreagh had previously delivered to *Adventure* magazine which were largely about his adventures while pursuing his various hunting trips. The change in focus now sees chapters on Ethiopia's cultural, political, religious and social history, as well as the aspirations and problems that this Christian country of ten million then faced.

MacCreagh also explained that part of the reason for writing his book was to encourage ordinary people to "go-look-see" and explore by showing how easy it was to make the trip and to do what he did. The secret, he stated, was in knowing 'the ropes' of travelling. However, as we shall see, things did not run smoothly, even for a veteran explorer like MacCreagh who already knew most of the ropes. He informs the reader that the total cost of this year-long trip for Helen and himself was $6,000 – a sum approaching $90,000 today. Hardly an amount that an ordinary citizen could easily afford, even in the wealthy United States of that time. His comment about the cost of the trip does however, give an indication of the author's relatively healthy financial position at the time. He was clearly earning a decent living from his adventure stories and royalties from his previous book, *White Waters and Black*.

Gordon and Helen MacCreagh Planning their Ethiopian Trip

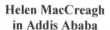

Helen MacCreagh in Addis Ababa

MacCreagh draws the reader into his Ethiopian story by explaining just how difficult it proved to be, in the great city of New York, to find out any information about the country and how to get there. After extensive enquiries, the best advice was that the quickest route was to sail to London, then cross the English Channel to Paris and down by train to Marseille. From there, a steamer would take the MacCreaghs through the Suez Canal and on to Djibouti in French Somaliland where, after a few days, they could take a train to finally arrive in Addis Ababa. However, once they actually arrived in Djibouti, they discovered that they could have taken a boat direct from New York to Aden and from there sailed on an overnight ferry to Djibouti – all at half the price of the journey they had just made. Since MacCreagh was personally funding much of the expedition's costs, this discovery clearly riled him. Perhaps out of pique, MacCreagh labels Djibouti as "the world's worst port".[59]

There was some entertainment to be had in Djibouti however, and the MacCreaghs hired a local guide to show them the town's night-life. They saw an Arab dance which consisted of two drums, a reed pipe and half a dozen men jumping up and down which quickly became boring. Then, a Somali dance which consisted of two drums, a reed pipe and half a dozen men jumping up and down and stamping their feet. This too soon became wearisome. Finally, they watched a Sudanese dance which also consisted of an equally boring two drums, a reed pipe and half a dozen men jumping up and down and stamping their feet. Needless to say, MacCreagh's opinion of Djibouti was not improved by this nocturnal sortie. He had one other comment to make about the port in his book, written in his typical wry style "It has been said there are twenty-seven trees in Djibouti. I didn't see all of them."

---

[59] MacCreagh states that at the time of writing his book, he had passed through the Suez Canal six times.

The couple's own feet had scarcely touched down in the dust of Addis Ababa when rumours of a man-eating hippo or *gumaré* as the locals called it, immediately pricked MacCreagh's interest. He quickly took the decision to become the self-appointed slayer of this local menace that lived in an area of rivers and lakes south of the capital. Within days, he had made arrangements for a mini expedition involving a six-day trek across barren, thorn bush-covered territory to find the offending beast. The fact that the monsoon season would soon start and the region into which he was heading had recently seen an armed insurrection by local tribes did not seem to overly concern him. MacCreagh soon managed to recruit an experienced guide to accompany the expedition, a British man called Jim, a hunter and ivory-poacher who had lived in East Africa and Ethiopia for many years.[60] He also hired some local men and their mules to porter the group's copious baggage. Not wishing to miss the adventure, his wife, Helen, "the lady who won't be frightened off", also insisted on coming along. However, Rossman, the expedition's cameraman, elected not to accompany the group into the bush and much to MacCreagh's disgust, remained behind in the city.

MacCreagh's account of his hippo hunt is very entertaining, if a trifle long. However, as we slowly trundle along with MacCreagh on his mule, across the dusty, black-brown earth, we gradually learn more about the man as well as interesting details about the country and its people. He states that was a non-smoker, moderate in his drinking habits, short-sighted and seems to have held no particularly strong religious beliefs, despite his grandfather being a church deacon.[61] MacCreagh's views on big game hunting and the killing of wild animals are in

---

[60] According to MacCreagh, Jim was also a gun-runner – see further comments in Chapter 11.

[61] If MacCreagh was indeed a non-smoker, as inferred in his book, other accounts of his later life clearly indicate that he smoked.

some ways, curiously ambivalent, though they probably reflect the general attitudes of the time. However, to his credit, he strongly disapproves of those hunters who set out to kill as much as they can just for trophies. Also, he must be one of the first European or American writers to publicly criticise the fur industry – "that beastly anomaly of civilization" as he refers to it in his book.

The author also gives his readers detailed insights into the way of life of the local tribespeople – the Nagadi who provided the mules and the Galla who live in the bush, surviving almost entirely on curdled milk and rancid butter. The author is often critical in his book of the natives' behaviour and customs but never in an overly condescending way. On occasions, the views expressed and terminology used by MacCreagh to describe the indigenous people might be criticised by some readers today as those of a bigot or racist. But that would be far from the truth. He writes honestly and openly with good humour about what he sees and thinks but his words are never those of a racist, especially taking into account the prevailing public attitudes in the 1920s. MacCreagh clearly showed an empathy for the native people he encountered on his travels, both in Ethiopia and elsewhere. When he is critical, as he often is, the same type of pithy comments could have equally been made by him about an over-charging cab driver in New York or an inefficient government official in France. He seeks to understand the situation of the local people he meets, is generally sympathetic to their problems and bears them no malice.

MacCreagh's description of the parched land through which the group travelled is vivid and entertaining. We learn something of the unexpected dangers of the trail. The "wait-a-bit", a green vine-like creeper with backward growing, small thorns. The plant's tendrils are like wire cable and the thorns strong as steel. If a passer-by should inadvertently be caught by a tendril, there's no way to pull free and escape; they have to stop and wait until they can cautiously unwind the thorny tendril up and back over their lacerated body. There are giant mosquitos that "stung clear through cord riding-breeches" and at night, growling leopards and raucous hyenas that constantly menace the camp. As the expedition nears the lake where the deadly hippopotamus resides, the group begins to encounter wild profusions of birds of many types and hues. There are eagles and hawks plus weaver birds that hang their nests upside down. "A playtime of pelicans" suddenly appear "like a squadron of white aeroplanes at five thousand feet, banking steeply to descend in wide spirals." Then there are the battalions of pink and white flamingos with "their stiff legs and their necks stretched ... they look like drunken tripods gone to sleep."

Strangely, for such an experienced traveller, MacCreagh seems to encounter a lot of problems and mishaps during the journey. But he relates these incidents in a humorous, self-deprecating manner, usually accompanied with a 'I should have known better' comment. On their very first day, Helen's mule bolted and dumped her into a mimosa thorn bush, leaving her badly scratched and with a seriously bruised leg. Then, one night, Hunter Jim's dogs got into the camp stores and ate their entire supply of bacon. On another occasion, as they made camp one evening, the Nagadi cook mistook Helen's precious remaining supply of cold cream for *Crisco* cooking oil and used it all up in preparing an interesting-tasting meal. No wonder that at times, the group's "tempers are like crackling matchwood." At one stage in his book, when things are not going according to plan, MacCreagh offers the excuse that "I have never trekked in Africa before." But of course, he had considerable experience, not only of similar conditions during his time in British East Africa but also of other expeditions and treks in arduous surroundings.

However, MacCreagh was a resourceful and persistent man who always manages to somehow come through the various difficulties that arise – in a similar fashion to the fictional Indiana Jones. It seems his wife, Helen also had ample resilience and pluck and MacCreagh is full of praise for her stoic endurance of the trials of the trail - the bolting mule, the loss of her face cream, the long hours in the saddle and her anxious concerns about the leopards and hyenas that prowled around their tent each night. He relates how on one dark, miserable night on the return journey to Addis Ababa, the couple became separated from the rest of their party and were lost in the bush without food or shelter. He relates how, in the midst of their frustration and despair, it was his wife who 'kept the vital spark alive', refusing to be overwhelmed by the situation. No wonder that MacCreagh referred to her as "the intrepid exploress" - she must have been a very special lady.

As the six-week, hippo hunting expedition drew to a close, after various entertaining and hair-raising episodes, the rainy season began. Previously dry gullies and river beds that had been easily crossed before, now became raging torrents, carrying the desiccated remains of long-dead dead animals along in their swirling waters. The expedition's mules constantly sank up to their knees in thick red-brown mud, called 'chicka' in Ethiopia. Making camp each evening was wearisome as they battled against the wind-driven rain. Lighting a decent fire to keep the hyenas at bay became a near-impossible task. The MacCreagh's return to Addis Ababa across a rain-soaked and very muddy country, turned into a painstakingly slow and miserable journey. They ran short of food, fought their way through swarms of flying termites that had risen up from the rain-softened ground and they became disoriented and seriously lost for a few days. They were a rather dejected and bedraggled couple by the time they arrived back in the Ethiopian capital.

The MacCreaghs spent the next few weeks exploring Addis Ababa while they waited for the weather to improve. Although MacCreagh found little to do in the city, especially during the rainy season, there is no doubt that he came across many things of great interest. For him, Ethiopia presented a thousand absorbing things to investigate. A fascinating melting-pot of people and races. A form of government that was basically still feudal. Slaves that refused to accept freedom. A ruler descended directly from the Queen of Sheba and the only African nation to have won a war against a European power and extracted an indemnity as a result.[62] MacCreagh's detailed comments about the Ethiopian people, their history and culture show that he clearly did a lot of background research. Indeed, he admits at one point to being unsure how to condense "the jumbled mass of material" he had collected.

When they weren't off exploring the bush, the MacCreaghs based themselves in the venerable, colonial-style Taitu Hotel in central Addis Ababa. It provided a fine panoramic view and with its verandas and shaded balconies, the hotel provided a welcome, cool retreat from the heat and noise of the city. It also seems that this is where MacCreagh gathered some of the information and rumours used in his book about the social and political situation in this part of Africa, both on this initial visit and during subsequent trips. He evidently spent many hours both talking with and covertly listening to the conversations of the hotel habitués - "the many old-timers who just naturally grow on hotel verandas."

In the second half of his book, MacCreagh richly details Ethiopia's cultural, political, religious and social history, as well as the aspirations and problems that this Christian country of ten million then faced. He also reveals a remarkable knowledge of world history and the political realities and diplomatic intrigues of that era. Some of his comments and conclusions were somewhat mis-guided or mis-judged and would today be regarded as dated or out of place. However, his assessment of the problems faced by Selassie and his government in trying to move the country out of the dark ages in just a few decades rather than the centuries it took Europe, were perceptive and generally sound.

He was extremely critical of the Europeans and Americans who complained about the state of affairs in

---

[62] MacCreagh was referring to the first Italo-Ethiopian war of 1895 – 1896 in which the better-equipped Italians were roundly defeated.

Ethiopia and often mis-represented the country in their homelands. He relates how one US journalist, who he describes as having "shirt-sleeve manners" had recently so upset the government that any American journalist was now a *persona non grata*. Given this context, it is to MacCreagh's credit that he managed to somehow develop a good relationship with Selassie and his officials. On the other hand, MacCreagh reveals that the odious reputation of American journalists (he was viewed by many as belonging to the "brotherhood of ink-slingers") precluded his hoped-for participation in a grandiose hunting expedition, then being organized by a titled, former British diplomat

Disappointed in not being able to take part, MacCreagh decided to arrange another expedition of his own into the interior. The rainy season was drawing to a close and he clearly had itchy feet. Based on a recommendation of one of his fellow hotel guests, MacCreagh set out for a secluded valley in the Ogaden region, near the Somaliland border, where there was reputedly plenty of good hunting. This safari was to be a much more modest affair than his previous one. Instead of mules, the MacCreaghs purchased two horses for their own use and hired a few camels to transport their equipment – although the camels turned out to be just as stupid and bad-tempered as the mules had been. Their earlier English guide, Jim, did not accompany them this time and his role was taken on by a local Amhara hunter who had proved himself on their previous expedition. Once again, Rossman, their cameraman, chose to remain behind, not wishing to risk his health in such a remote region. A few native camp boys completed the expectant hunting party.

Earl Rossman, Expedition Cameraman

Taitu Hotel, Addis Ababa

MacCreagh Setting Up Camp, Ethiopia

Helen MacCreagh with Leopard Cubs

Much to MacCreagh's annoyance and his wife's disappointment, this second hunting expedition turned out to be a complete waste of time and money. Initially, there was hardly any game to be found, despite several days of diligent searching. MacCreagh managed to shoot a great bustard and they did see a rhinoceros in the far distance, across a lake but it had gone by the time they reached the spot. However, as usual with MacCreagh, there are various escapades and adventures that take place along the way and he entertains the reader well by relating these in his typical, humorous style. As their small caravan began its return journey, the camel boy disappeared one night with all the camels, soon to be followed by the camp boys. Together with their Amhara guide, the MacCreaghs eventually managed to hire some donkeys and local men to transport the remnants of the expedition back to civilisation. Having painstakingly loaded up their new caravan, the MacCreagh's incredibly quickly lost touch with their hired donkeys and men, forcing the couple to make the trip back to Addis Ababa alone. I suspect that Earl Rossman was full of *schadenfreude* on their return which probably infuriated MacCreagh even further.

The author summarizes this chapter as "nothing but the failures of the heroic expeditionists." Perhaps a little unfair but true. However, to MacCreagh's credit, there is a forthright honesty in his writing which makes his account all the more readable. He doesn't shy away from relating his mishaps and freely admits when he has made a mistake or been at fault. His deference towards (and evident care for) his wife is also clear from the narrative. MacCreagh is no gung-ho hunter, dragging his reluctant wife along on his chaotic adventures. She is very much a partner in their enterprise and it is clear that the pair had a very close relationship.

As the MacCreaghs were making their way back to the Ethiopian capital, they heard news that the titled Englishman's expedition that MacCreagh so desperately wanted to join earlier had also run into trouble. It seems that the wealthy hunters had themselves been hunted, having been attacked by Somali tribesmen in the Ogaden and several members of the party had been killed. MacCreagh was excited by this news. The adventurous, go-look-see trait in his character took over and he was determined to make his way to where the attack had taken place to see what was going on. He hoped to garner some interesting, additional material for his book and probably for a newspaper article as well. Helen declined to return to the bush so soon and reluctantly agreed to her husband making the journey without her.

Once again, MacCreagh's account of his trip makes for absorbing reading, not because he is successful in reaching the site of the incident but for the characters and events that he encounters along the way. He is helped (and sometimes not helped) by various people during his trip. At one stage, a friendly Galla tribesman supplied him with five donkeys and two sturdy women as transport. As MacCreagh humorously remarked, although women can't carry as heavy a load as a donkey, "they make less noise than camels and cover more ground because they don't stop to eat cactuses." On a more serious note, the local Somalis came in for particularly vitriolic criticism. MacCreagh found them to be insolent, lazy and smelly. "The Somali in free Abyssinia is a gratuitous affront to all men, white and black. Never have I more wanted to shoot three men in the stomach all at once." According to the author, the only good point possessed by the Somalis was their undoubted courage.

As things finally turned out, MacCreagh was unable to continue into the Ogaden – the locals advised him it was simply too dangerous and volatile to travel there. However, he did manage to discover some background to the events that led to the deadly attack on the English nobleman's caravan. Inevitably, there were differing versions – the Ethiopian government's report, the local Ethiopian official's account and that of the British government.[63] Basically, it all came down to a tragi-comic lack of communication but that hardly appeased an indignant Britain, which demanded compensation. For MacCreagh, Britain's high-handed approach to the affair was evidence of its long-held ambition to add Ethiopia to its sphere of influence in East Africa. There was certainly a widely-held view among the European ex-pats in Addis Ababa that Ethiopia was a backward and poorly controlled country which would benefit from coming under the 'protective' wing of a colonial power. But MacCreagh's conclusions were based more on remarks made by the local veranda habitués than official British government policy. As we shall in the next chapter, it was Italy rather than Britain whose ambitions would turn out to be aggressive and acquisitive in regards to Ethiopia.

For any reader of *The Last of Free Africa* who is familiar with MacCreagh's earlier writings, there will be no surprise at finding a whole chapter devoted to witchcraft, wizards and werewolves. Alongside his constant interest in the habits, languages and customs of the local people he encountered on his worldwide travels, MacCreagh also regularly delved into their mysterious, mythical and magical legends. Although often sceptical of the claims made, he sometimes concludes his re-telling of these tales by conceding that they may well hold more than a grain of truth. It seems strange that an educated and worldly-wise man like MacCreagh not only took such an interest in these tribal myths but also occasionally credited them with an element of credibility. This apparent acceptance of supernatural or magical powers was an unusual trait in MacCreagh's character.

He also spends time delving into the domestic arrangements of Ethiopian households as well as the position of women in the country's society at the time.

---

[63] The British government subsequently set up a commission of enquiry into the incident that imposed an indemnity of several thousand pounds on the Ethiopian government for its dereliction in contributing to the deaths of members of the expedition.

This is a subject that clearly held great interest for him and once again shows the author's fascination with the lives of the ordinary people that he encounters on his various travels. MacCreagh was apparently both surprised and impressed with the status of women and their freedom in Ethiopia, particularly when it came to marriage and divorce. It was evidently, not what he had expected to find in an African country. The women in Ethiopia maintained complete independence in their own affairs, careers and homes. According to MacCreagh, "it is the women who have framed the conventions of marital conduct" whereas, in most countries of the world, the rules and regulations have been arranged by men. He sees Ethiopia as a country of easy marriage and easy divorce – at least from the woman's viewpoint – since they can choose whom they marry and how or when to divorce. MacCreagh describes in some detail the different forms of marriage, the traditions and ceremonies involved as well as methods for achieving a divorce or dissolution.

These comments in his book clearly aroused interest and some indignant comment back home. For example, on their return to the USA, an article headed 'Wreaths and Wallops for Women' was printed in the Kansas *Atchison Daily Globe* on October 15th 1928. The female columnist writes about the views expressed by Helen and Gordon MacCreagh on how well a woman's place in Ethiopia compared with the situation in the USA. That Ethiopian "women have in every way more equal rights than the men . . . and the law backs them up" seems to surprise the rather sceptical Kansas reporter. She ends up by writing "where do the MacCreagh's get the idea that the women in Ethiopia are any better off than our American women."

Sadly, there is not much mention of the search for the Ark of the Covenant until towards the end of the book. Even then, MacCreagh has little to say about the matter which is disappointing, given it was one of the main original objectives of the trip. One is left with the impression that it was almost forgotten about as other matters, especially hunting, attracted his attention and took precedence. It is also likely that the realities of travelling across Ethiopia to search for the Ark during the monsoon season played a part in his decision-making – a reality-check that Indiana Jones didn't have to face in his cinematic search. In a chapter solely devoted to the history of Christianity in Ethiopia, MacCreagh considers the history of the Ark and how it potentially came to be located in the country. He accepts as plausible, the Ethiopian Orthodox Church's view that it is housed in an ancient temple at Axum, in the far north of the country.[64] Similarly, the same chapter only gives the Falashas a brief mention. There is no indication of a trip by MacCreagh to visit them as originally planned when he and his wife set out from New York.

Although in the end, MacCreagh failed to visit both the Ark of the Covenant and the Falasha tribes, he did manage to travel quite widely in Ethiopia during his stay. This was a quite remarkable achievement since travel by foreigners outside Addis Ababa normally required permission or letters of introduction to the local governors. From the comments in his book, it is unclear whether MacCreagh simply ignored these regulations in his typical gung-ho fashion or was in fact given permission to freely roam the country by the Ethiopian government officials he had met and befriended. If it was the latter case, it is possible that this was because he was an American, the only major power that did not covet the rich, undeveloped resources this isolated, land-locked nation in the horn of Africa had to offer.

In his book's final chapter, MacCreagh discusses in some detail the political and economic problems then facing "the last free country in Africa". He reviews the potential threats to Ethiopia's continued independence,

---

[64] Axum is one of the oldest, continuously occupied towns in Africa, founded in the fourth century BC. The Ark of the Covenant is reputedly held in Our Lady Mary of Zion church there.

coming mainly from Italy and France who, the author alleges, coveted her agricultural and mineral riches. MacCreagh readily admits that he has no proof that Ethiopia's "neighbours wish to gobble her up." Instead he merely presents what he refers to as "pertinent questions" about their true ambitions or motives and the impact these were having on Ethiopia at that time. However, the innuendos in his book about European colonial ambitions were probably not well received by the relevant governments nor indeed by many of the Addis Ababa veranda habitués. In his previously mentioned 1961 article, Bob McKnight wrote that MacCreagh's book was "a prophetic treatment in view of what has taken place since."

The Last of Free Africa can be read on several levels: as a travel book, a history book filled with social commentary or as book of entertaining and thrilling adventures in the wild interior. At the time of publication, it might indeed also have served as a 'how not to' guide for any contemporary hunters or adventurers contemplating a similar overseas trip. Overall, it's hard not to conclude that MacCreagh was more intent on entertaining the reader than making a serious contribution to scientific knowledge about Ethiopia. As good and enjoyable as the narrative of MacCreagh's book is, it's his fascinating photographs (and their often-amusing captions) that add an extra dimension to the narrative for the reader. There are some ninety black and white illustrations spread through the book which give an excellent insight into both what MacCreagh saw and did as well as what daily life was like for the local people at that time. Ethiopia fascinated MacCreagh and his interest in the country and its people is evident in every chapter.

Chapter Nine

Ethiopia Twice

After his return to New York, MacCreagh spent the next few months writing his book. As he was preparing it for publication in the summer of 1928, a unique opportunity arose to visit Ethiopia once more. Too good a chance to miss, MacCreagh grabbed it with both hands. This time, he was recruited to be the leader and guide for what became known as the Sanford-Legendre expedition. The Department of Mammals of the American Museum of Natural History in New York was keen to acquire additional animal exhibits for a new African Hall in the museum and with the help of sponsors, it arranged an expedition to Ethiopia. The primary sponsor was John Sanford, who was heir to the large New York carpet manufacturing company Bigelow Sanford and a member of the U.S. House of Representatives. He partly funded the expedition as well as subsequently promising to provide money for the mounting of the exhibits.[65] Apart from MacCreagh, the other members of the expedition were Donald Carter, the museum's curator, John Sanford's daughter, Gertrude, Sidney Legendre and his brother Morris. In addition, Sidney and Morris Legendre agreed to provide financial assistance to help cover the expedition's travelling expenses. According to Peter Ruber's account of MacCreagh's life, his wife Helen also went on this trip. Given her love of adventure plus the chance to travel with such an interesting group of people, this appears highly likely to be correct.

---

[65] According to the American Museum of Natural History's May 1929 report, the sum paid by John Sanford was $25,000.

Morris and Sidney Legendre, Ethiopia

The Legendre Brothers and Gertrude Sanford

Rather similar to the Mulford expedition to the Amazon Basin several years before, the Sanford Legendre trip also turned out to be a high-profile affair. Both the Sanford and the Legendre families were part of American 'High Society' with their activities regularly featured in the gossip columns of the US press. Gertrude Sanford in particular was well-known to the public. Her mother, Ethel, came from a family of international diplomats, one of whom was an associate of Abraham Lincoln and the founder of the town of Sanford, Florida. Born in 1902, Gertrude was, despite her wealthy upbringing and charmed life, as much at ease in hunting garb staring down the barrel of a rifle as she was in high heels. The expedition to Ethiopia with MacCreagh was not her first hunting trip. She was in her teens when she travelled to the Grand Tetons in Wyoming and shot her first elk and in 1927 she had visited East Africa, where she killed her first lion. The Legendre brothers hailed from a wealthy, aristocratic family in New Orleans and both were keen hunters and accomplished horsemen, who enjoyed travelling and adventure.

We don't know how MacCreagh became involved with this particular expedition and its high-profile, celebrity members but he was presumably recruited due to his experience of big-game hunting and the back-country of Ethiopia. The group arrived in November 1928 and spent the following nine months shooting a large collection of Nyala antelope,[66] several tigers as well as many smaller mammals plus various birds and reptiles. The expedition ran smoothly and MacCreagh was evidently able to guide his employers to good hunting grounds. A very different experience from his own visit to Ethiopia a year earlier. Most of the collection was destined for the museum in New York but some specimens were kept as trophies by Gertrude and the Legendre brothers. Given MacCreagh's

---

[66] A rare, highly-prized, mountain antelope found only in the highlands of Ethiopia.

views on such trophy hunting, he presumably had to swallow his pride and turn a blind eye. It turned out that during the expedition, another kind of hunting was also taking place, as Gertrude became infatuated with and subsequently bagged handsome young Sidney Legendre. On their return to the USA in 1929, her father was faced with not only making good on his earlier promise to fund the mounting of the considerable collection they had brought back for the museum but also the unexpected cost of a high society wedding.[67]

With their shared love of hunting and roughing it on safaris, the newly-weds later continued their overseas travels with a trip to French Indochina (now Vietnam, Cambodia, and Laos) to acquire more exotic creatures for the museum's collection. In subsequent years, they travelled through Southwest Africa and Iran. When World War II erupted, Sidney signed up while Gertie volunteered, eventually working in London for the Office of Strategic Services, the forerunner of the CIA. After the retaking of Paris by the Allies in 1944, Gertrude was transferred to the city. A few weeks later, during a car journey to the front line near the German border in Luxembourg, Gertrude and her companions were captured by the Nazis. She was subsequently held as a prisoner of war for six months until she managed to escape on a train bound for Switzerland. When the train stopped for inspection short of the border; she leapt out and dashed towards the frontier. A German guard ordered her to halt or be shot but she bravely continued and safely crossed the border into Switzerland.

Soon after MacCreagh's arrival in Ethiopia with the Sanford-Legendre expedition, he was awarded the Order of the Star of Ethiopia in recognition of his services to the Emperor and his country. The reasons for this prestigious honour have never been fully made public and are

---

[67] In 1930, Gertrude and Sidney Legendre published their own account of the trip to Ethiopia – *In Quest of the Queen of Sheba's Antelope*.

therefore subject to speculation. Typical of the press reports of the award was that in the *Charleston Gazette*, published on November 11th 1928:

"Gordon MacCreagh ... has been awarded the Order of the Star of Ethiopia in recognition of his services to the King of Abyssinia. Although the facts of the case are necessarily concealed, it is known that Mr. MacCreagh brought to the attention of his Imperial Highness, Tafari Makonnen, King of Abyssinia, a piece of International intrigue of several European nations, which If it had gone through would have greatly prejudiced the economic life of the country."

How this honour came about was never discussed by MacCreagh, either in his *The Last of Free Africa* or in subsequent interviews with the press and it's another of those mysteries relating to MacCreagh's life. Certainly, the author was very complimentary in his book towards the Emperor and he showed a sympathetic understanding of the country's history and problems. But this hardly seems reason for the award of such a prestigious and unusual honour, which occurred just a few weeks after the book was published in America. That hardly gave enough time for it to be shipped to Ethiopia, its contents read and digested and an award to then be made. MacCreagh later claimed in another press interview that he was the first white man to receive the award but this was not correct. This may have been a genuine error on his part of simply another example of him embellishing reality in an off-hand remark.

Some biographical sources claim that MacCreagh became a close friend and advisor to Emperor Haile Selassie. [68] While this may well be true (as we shall see later), it is far from clear whether MacCreagh ever actually met Selassie prior to receiving his exceptional honour. There is no indication of any meeting or discussions

---

[68] Both Peter Ruber and Bob McKnight mention this in their accounts of MacCreagh's life as well as several US press articles.

between the two men. Also, the photographs of Selassie in MacCreagh's book appear to be generic and don't actually show the two men together. During the few months that MacCreagh spent in the country on his first visit, he was mostly away from the capital, hunting in the bush. His relatively limited time in Addis Ababa would not seem to have been sufficient for a close relationship to have developed, especially as MacCreagh was initially regarded as one of those disreputable American "ink-slingers".

Similarly, none of the many press articles that covered his initial visit to Ethiopia, the reviews of his book or indeed the subsequent award, give any clue as to whether the two men met – with one exception. In a report published by the *Laredo Daily Times* of Texas on April 21st 1927, covering the recent arrival of MacCreagh's first expedition in Addis Ababa, it states he was given an audience with Haile Selassie and received with full splendour and ceremony. The newspaper goes into some detail about the visit and MacCreagh's discussions with Selassie as well as his observations on the sumptuous magnificence of the imperial palace. According to the article, the two men had a lengthy conversation about a wide variety of subjects and MacCreagh was impressed that Selassie was fully up to date with world events as well as recent technological developments. Given the detailed nature of the contents, it is hard to ignore or disregard what it says. However, it is odd that these comments were neither reflected in MacCreagh's subsequent book nor in any other press report.

There is no doubting that the remarks by MacCreagh in his book about the diplomatic intrigues and political threats to Ethiopia at the time of his visit were both insightful and revealing – at least to most of his readers. But it hardly seems credible that his 'revelations' were news to the Ethiopian government. The rumours and European newspaper reports that were at the heart of MacCreagh's comments would surely have been picked up by the administration. Neither does it seem likely that while in Addis Ababa, MacCreagh came across some highly secret information that he was able to pass on confidentially to Selassie's government. Virtually all the comments and allegations made by MacCreagh in his book were matters that had been well discussed in the European press. As he himself said, the Ethiopians did not need "the prompting of the gentlemen of the cafés" to know what was going on. His unusual reticence about the honour suggests there may have been another, additional explanation for the award. One based more on the actions of the author rather than the words that he wrote – see later comments in Chapter 11.

Part of the reason for the appeal of *The Last of Free Africa* during the 1930s, at least among the general public, was the fact that MacCreagh provided a refreshingly different viewpoint on the country and its people. It was one that seemed to strike a chord with his readers, especially those in America. Although other books and press articles about Ethiopia were relatively rare, those that were published proved mostly critical, representing the country as a very backward place, desperately in need of modernisation and change. Typical of these was the famous British writer, Evelyn Waugh who in October, 1930, visited Ethiopia to cover the coronation of Haile Selassie for several major newspapers. He reported the event as being "an elaborate propaganda effort", organized to convince the world that Abyssinia was a civilized nation that concealed that the Emperor had achieved power through barbarous means. Waugh returned to Abyssinia in August 1935 to report the opening stages of the Italo-Ethiopian War for the *Daily Mail*. According to one of his fellow reporters, Waugh considered Abyssinia "a savage place which Mussolini was doing well to tame". [69] In his own book on Abyssinia, published in 1936, Waugh wrote, "The Abyssinians, in spite of being by any possible

---

[69] William Deedes, later became a Conservative politician and Editor of *The Daily Telegraph*.

standard an inferior race, persisted in behaving as superiors; it was not that they were hostile, but contemptuous." Waugh left Addis Ababa in December 1935 with great relief, a feeling no doubt shared by the locals.

In stark contrast to the bigoted views expressed by Waugh, was an article by MacCreagh published in 1935 in the *Saturday Evening Post* entitled 'King of Kings'. According to a subsequent review of the article, MacCreagh provided "an insightful glimpse of 1930s Ethiopia. Through his visit to Ethiopia, he thoroughly praises Emperor Haile Selassie, whom he regards as exceptionally wise and virtuous. Conversely, he shows Western animosity and disgust to much of the rest of Ethiopia's culture, geography and indigenous groups."

The reviewer goes on to explain how MacCreagh's article also discusses "the concept of blackness and associates it as disingenuous with values of civilization and propriety. These values of high culture are in his viewpoint, inexplicable and have no place in a Black or 'backward' continent. This can be further explained by the prevailing 'Dark Continent' narrative. Through his proceeding observations, he wonders how can such a noble ruler be African? He also makes an interesting parallel to the Ethiopian's own sense of distinctiveness and self-superiority. As he defines their sense of pride: "Black Aryans," there is an expectation of victory, religious righteousness and delusion."

The insight, empathy and balanced discussion in MacCreagh's article were a refreshing counterpoint to the views expressed at that time by Waugh and much of the biased western colonial press.

The period between 1927 and 1930 must have been an extremely busy and exciting time for MacCreagh. The visits to Ethiopia, the publication of his new book and the award of the Star of Ethiopia were no doubt the highlights. Although his fictional output understandably declined for a while, he still managed to publish three adventure stories in 1927, two more in 1928 and again in 1930. In addition, he was also heavily engaged in giving lectures across the country to promote his book as well as to draw the attention of the American public to the problems in Ethiopia. MacCreagh also found time in his busy schedule to participate in a couple of radio shows. The first was broadcast from the New York Exchange Club in October 1929 and the second in June 1930 by NBC. Both programmes essentially featured talks describing the most exciting moments of the speakers' lives. It would be interesting to know which event was chosen by MacCreagh – he had so many to choose from.

At some stage around 1930, MacCreagh made a small donation to the American Numismatic Society in New York that included two blocks of salt (*amole*), and an assortment of coins he picked up in Ethiopia plus two corroded gun cartridges, with their jackets intact and presumably filled with powder. The bullets are 9mm and were produced in France, as indicated by the stamp S. F. M. (Société Française des Munitions) on the bottom of the casing. They served as ammunition for the Fusil Gras, a popular and fast-firing French service rifle that was manufactured in volume during the late nineteenth century. The cartridges and *amole* were examples of popular forms of commodity currency in Ethiopia at the time of his visits. When making his donation, MacCreagh mentioned that a single bullet was equivalent to 1/5th of a Maria Theresa thaler, the thaler still being a popular currency in that part of Africa. Apparently, five cartridges would purchase a goat or several chickens. However, the subsequent war against the Italians saw such a massive increase in the availability of arms and ammunition such that the value of each cartridge plummeted dramatically.

Chapter Ten

Third Class Travel

Although MacCreagh's travel and writing focus now seemed to be firmly on Africa, he continued to retain an interest in India and Burma. He was one of the early members of the Kipling Society, founded in Britain in 1927 to celebrate the life and works of Rudyard Kipling. It would seem that MacCreagh joined the Society that year, presumably during his visit to London whilst on route to Ethiopia. Kipling was considerably older than MacCreagh but the two authors shared a common affection for tales set in Imperial India and they may possibly have met and exchanged stories at one of the Society's early meetings. Coincidentally, Kipling was also the honorary president of The Ends of the Earth Club in New York which was similar in scope to the Adventurers Club. MacCreagh was a member of both clubs. The former held an annual dinner at the Savoy Hotel in New York so it is highly likely that the two men spent time together there.[70] Certainly, from comments made in *The Last of Free Africa*, it seems certain that MacCreagh was very familiar with both Kipling and his writing.

Following his two visits to Ethiopia in the late 1920s, MacCreagh decided to return to East Africa sometime in late 1928 or early 1929. Once again, Helen, just as eager as her husband for new adventures, accompanied him. MacCreagh later wrote about this decision by his wife:

"Fate had descended upon me...and she was crazy, too, and came along -- and suffered for her temerity...Thin tent walls out in the open bush and lion noises outside rasped her nerves all up. And drove her crazier; so she came out again the next time."

---

[70] Mark Twain was also an early member of The Ends of the Earth Club.

That next time was a journey that would take them further into the interior of the British East Africa and across the Uganda border. Once again, this trip was to be a hunting expedition and on arrival in Kenya, MacCreagh made arrangements for transporting their baggage up-country. Wary of the problems involved in hiring animals and their native handlers from his first visit to Ethiopia, he decided he would purchase his own mules. As MacCreagh later described it, the ones he bought had been "scientifically inoculated" locally against the tsetse fly.[71] However, despite the vendor's claims about the animals' immunity, the tsetse flies succeeded in killing off half the mules during the first part of the journey into the interior. And then, as the MacCreaghs struggled further on into the wild bush-country, the lions ate up the rest. So MacCreagh and his wife were forced, much against their will, to pay for a local safari outfit to help them to continue with their journey. But apparently, since the safari's own mules were ridden with tsetse, many of them died or were unable to carry their loads. Then, the porters soon ran away in droves taking most of the MacCreagh's luggage and supplies with them. The story is almost an exact repeat of what happened on their first Ethiopian trip some six years earlier.

Despite MacCreagh's careful and sensible planning, the couple now found themselves abandoned and alone, stranded in the remote savanna. With no shelter and hardly any food or medical supplies, they both became increasingly ill. As they struggled on, trying to retrace their steps, the rainy season started, further adding to their woes. After a couple of extremely difficult weeks, they somehow managed to reach civilization on the coast of British East Africa. Once they had recovered somewhat from their trouble-filled journey, they were able to book

---

[71] In his book, *The Last of Free* Africa, MacCreagh mentioned inoculation experiments on mules that were then being conducted in East Africa against the tsetse fly and queried their efficacy. Perhaps he should have known better this time.

accommodation on a French ship that was sailing from Mombasa to Japan. With their money running low, the MacCreagh's could only afford third-class accommodation on the French vessel. As MacCreagh later wryly remarked "don't you ever try that." From Japan, the couple were eventually able to find a ship bound for Seattle and they finally landed back in the USA in July 1929.

Having finally made it back to the USA from their exploits in East Africa and a tortuous journey across the Pacific, one might have thought that the MacCreaghs would have taken it easy for a while. But the pair's indomitable yearning for adventure and travel literally drove them on to further exploits. In Seattle, they purchased what MacCreagh later described as "a used - very used flivver" and then proceeded to drive right across the USA.[72] The couple travelled overland from west to east, stopping at many of the country's famous scenic wonders and historical sites. It was almost three months before the couple finally arrived home in New York.

In MacCreagh's short and rather cryptic, autobiographical account published in the *Argosy* magazine a few years later in February 1933, (see Appendix I), he provides a brief description of their trip.

"We took in [the] Columbia River Highway and Yellowstone Park and Jackson's Hole and Shoshone Canyon and Deadwood and Cody and Custer, Dead Man's Gulch and Two Mile Bend and Snake River and Massacre Rock and Poison Springs. And we got an awful kick out of it all."

Following their return home, MacCreagh again picked up his writing career, publishing two or three stories yearly for the next decade. Despite the many problems he had experienced during his visits to Africa in the late 1920s, they clearly gave him inspiration for new tales of adventure. He began writing what would prove to be some of his most successful and enduringly popular stories.

---

[72] US slang for a small, cheap and usually dilapidated, old car.

Known as the "Kingi Bwana" tales, they are set in Africa and feature the adventures of an interesting American character named King who the natives refer to as "Kingi Bwana". The first story to feature him was *The Slave Runner* which was published in the April 1st 1930 issue of Adventure magazine, quickly followed by *The Ebony Juju* a couple of months later. In these tales, King is described as a rugged Westerner, having grown up in the USA's Dakota Territory. His day job was running safaris but he appears to spend most of his time as a local troubleshooter. He is portrayed as a rather shady character, often operating on the wrong side of the law as a slave runner and a smuggler but of course, in colonial Africa, things were not always as they seemed. As the story unfolds, we learn that "Anything can happen in Africa!"

MacCreagh's writing style can seem a little stilted or old-fashioned at times when viewed by today's standards but he is inventive with his plots and generally holds the story together well, keeping it moving forward at a decent pace. As one might expect with MacCreagh, *The Slave Runner* story is enhanced by the considerable detail included about Africa and its geography, politics, wildlife and social customs as well as the attitudes of its people. There is a good balance between the protagonist, King, and two British government officials, a pompous deputy commissioner called Sanford and a young, earnest consul. The former captures the American and accuses him of slave running because King is always in the same vicinity as a notorious Arab slave trader. The suspicious deputy commissioner is convinced the two men are partners in the illicit enterprise.

Many of the characters and events are clearly derived from MacCreagh's own real-life experiences. King's personality and activities seem to be based on a mixture of those of Alfred Klein, the safari operator MacCreagh met as a young man in Nairobi and 'Hunter Jim', the guide that he hired on his first trip to Ethiopia. Similarly, the two British government officials were probably based on real people that MacCreagh encountered as a young man while in British East Africa. There's a long, suspenseful scene where King is penned up in a lion trap, only to have the lion he was hoping to capture come along and try to get to him. King's escape from both the trap and the lion makes for some entertaining reading. This part of the tale was partly based on an actual event in MacCreagh's own life some years before when he was in Burma. He would also use the same idea for a story he wrote for *The Literary Digest* a couple of years later (see below).

In *The Ebony Juju*, published in July, 1930, King is asked by a British government official to investigate rumours of gun-smuggling and a possible native uprising around Lake Victoria in Africa. After nearly being killed by the gun-runners, King eventually locates the extremely restless natives who are being stirred up by evil witch-doctors using a large idol carved of ebony that can move and talk when it's possessed by the spirits. This is a familiar scenario, having featured in at least one Tarzan movie as well as having echoes of scenes from the Indiana Jones film scripts set in India and Peru. Predictably, but not without many difficulties, King finally sorts everything out and thwarts the dastardly villains

In January 1931, MacCreagh's third story in the Kingi Bwana series was published by the Adventure magazine. Intriguingly entitled *The Lost End of Nowhere*, it was almost 38,000 words long – it could almost have been a novel. Once again, the tale has links to MacCreagh's earlier life as King is contacted by a German university asking him to locate a scientist who disappeared in what was then German East Africa fifteen years earlier. King decides to take on the search, and along with his perennial loyal companions, the fierce Masai warrior named Barounggo and the wily Hottentot known as Kaffa, he sets out to find the lost scientist. While they work their way through the jungle, the trio battle with unfriendly natives and come across various mysterious rumours about the missing scientist. As MacCreagh's story reaches its climax, there's plenty of typical pulp fiction adventure and authentic blood and thunder action before King saves the world. If Hollywood ever decides to make an Indiana Jones movie set in East Africa, then the Kingi Bwana stories would make an excellent basis for the script.

As with his book about Ethiopia, *The Last of Free Africa*, MacCreagh's Kingi Bwana tales contain various comments that appear to express the author's own views on social and environmental issues of the day. There is a definite empathy with the local native people and although King is clearly the leader, he and his loyal sidekicks are portrayed as brother adventurers under the skin. There is concern too for the indigenous wildlife. MacCreagh strongly believed in only killing wildlife to protect one's life or to provide food. He disapproved of those who shot big game simply to have photos taken with their quarry or to have them stuffed for display back home. Kingi Bwana's outspoken criticism of the policies of colonial governments also echoes MacCreagh statements about European colonial ambitions concerning Ethiopia in The Last of Free Africa. In many ways, King comes across as the fictional embodiment of MacCreagh, a man who followed his own instincts without regard for the dangers that lay ahead, and who strongly believed in the principals described above. These are not just ordinary tales of adventure and mystery, they are good stories with a message that continue to be read and enjoyed today.[73]

Following the success of his initial Kingi Bwana tales, MacCreagh ventured into a new form of fiction writing with the first of his two Dr Muncing novelettes. This was initially published in the launch issue of *Strange Tales of Mystery and Terror* in September 1931. The magazine was brought out as a rival to the established *Weird Tales* to which MacCreagh would also later contribute. The unusual, eponymous hero is a doctor-detective who seems

---

[73] For example, see James Reasoner, *Rough Edges* blog from 2017 and 2018.

to have clairvoyant powers that he uses to investigate mysterious paranormal or psychic events. This was followed a few months later by *The Case of the Sinister Shape* which appeared in the same magazine. The fact that it only ran for seven issues and folded in early 1932 probably explains why MacCreagh didn't produce any further Dr Muncing episodes and returned to his more normal adventure stories.

In early 1933, MacCreagh is recorded as living in Centerport, a small but fairly affluent village on the north shore of Long Island, where he continued to pursue his writing career.[74] Soon after he moved to Centerport, he met his new neighbours in a rather unusual and unexpected manner. One evening, they heard MacCreagh outside in their yard calling out "here, pussy, pussy" several times. When they went out to see what was happening, MacCreagh explained that he thought his pet cat might have got into their cellar. The neighbours fetched a torch and together with MacCreagh, crawled into the small cellar where his 'pet cat' was quickly discovered. Staring back at the trio in the dark were a pair of glowing eyes belonging to a half-grown leopard with paws as big as a man's forearms. MacCreagh's neighbours fled in terror, leaving him to retrieve his so-called pet. He later explained that he had brought the animal back from Africa and claimed it was fully domesticated and not at all dangerous. MacCreagh's nervous neighbours remained unconvinced, especially when they learnt it had broken free from a chain of one-inch thick steel rings. Soon after, MacCreagh wisely donated the leopard to the local zoo.

---

[74] *Mattoon Gazette* article of January 1933.

Adventure Magazine Front Cover, 1930

Argosy Magazine Front Cover, 1934

Following their initial, difficult meeting, the MacCreagh's became good friends with the couple next door, remaining close until they moved to Florida. The details of what happened that night appeared several years later in the *Key West Citizen* of August 4th, 1940, written by the neighbours who were now also resident in Florida and contributors to the newspaper. By pure coincidence, they had bumped into MacCreagh one day in Key West and the chance meeting reminded them of the earlier incident which they wrote up as a story for their newspaper. Intriguingly, in both this article and a subsequent one written about MacCreagh, they describe him as being British, not American. They mention that one of his favourite words was 'incredible' which they felt was a very English expression. Given that they knew MacCreagh well, their assertion about his origins is significant and highly credible.

In August 1933, MacCreagh entered a competition run by *The Literary Digest* for stories about the 'Narrowest Escape from Death'. His short story, with the title of *Trapped by a Man-Eater*, described the challenges and dangers involved in trying to capture a man-eating tiger in Burma and it won first prize. It follows the same basic 'true-life' scenario MacCreagh used in his earlier *Slave Runner* tale. The story is also reminiscent of one of his exploits in Ethiopia when he set out to shoot a marauding hippo that was threatening local tribespeople. His frequent use of these life-threatening situations while dealing with wild animals in his writing indicates he probably also employed the same theme to entertain his lecture audiences or even friends invited round to his home for dinner.

The man-eating tiger, described by the local Burmese as "a witch tiger," and having "the soul of a sorcerer" was a menace to their village and had already killed three people. MacCreagh was determined to catch it and set up a large trap with a live goat tethered inside as bait for the tiger. MacCreagh built a trap, shaped like an A frame, with a falling trap-door released by a catch that the tiger would have to tread on to reach the live goat pinned at the far end as bait. He sat and waited nearby with his gun and although the tiger came night after night, it was too wary to venture in.

So, the next night, MacCreagh relocated himself and constructed a basic platform about ten feet off the ground in an adjacent tree. "Clammy hours passed," wrote MacCreagh. "Queer things whispered in the underbrush. Shadows made shuffling noises. Then all of a sudden it was there. If I saw what I thought I saw, it must be an enormous brute. Something looming dimly motionless before the door of the trap. Hungry. Whining in throaty indecision. A monster." At the very moment when the tiger seemed poised to enter the trap, MacCreagh's platform collapsed and he unexpectedly found himself "in the same small jungle glade with the killer, on his own ground, in his own night." He had lost his rifle when he fell from the tree and could not waste time groping around for it. He was almost face to face with the tiger when it gave a loud, coughing roar. There was only one safe refuge now for MacCreagh – inside his own trap just a few feet away.

Full of terror, he made a hurtling dive into the trap and as he squirmed into the narrow space alongside the goat, he accidentally tripped the trap-door lever and it slammed down shut behind him. "And on the very heels of the slam came the grating snarl of the killer." For the rest of that night, MacCreagh was effectively held captive by the huge, man-eating tiger. As MacCreagh crouched terrified inside the trap, the animal constantly prowled around it, looking for a way in. "Immense eyes blazed between the close-set poles of the trap," he wrote. "They blazed and blinked and went out—and suddenly glared startlingly again from another place. Hot breath snuffled in at me. Claws rasped along my wooden grille. Sometimes the whole structure would shake." Eventually, with the approach of dawn, the frustrated tiger disappeared into the darkness. MacCreagh never got his tiger, but shortly afterwards, it was shot by a local tribesman.

However, *The Literary Digest* got an entertaining story, MacCreagh a prize, and presumably, the goat a reasonable hope of a longer life.

In the same year of 1933, MacCreagh won another writing competition prize – this time a new Chevrolet car for an essay on why he liked his Chevrolet. His winning entry was based on a tale from one of his visits to Ethiopia. He described how a crazy Arab driver, who had just graduated from camels, drove into a local man and killed him. The Arab was arrested, tried and eventually publicly hanged as a warning to other drivers. "And a damn good law it is too," MacCreagh concluded. He was evidently keen on competitions, or at least those that involved photography or some creative writing. In September 1938, he won first prize in a photographic contest in St Petersburg, Florida with a picture of the city's Cypress Gardens. Then, in July 1940, the mayor of St Petersburg ran a contest to find a new safety slogan for the town, with entries split into two categories – drivers and pedestrians. MacCreagh entered and although he didn't win, he was cited as one of the runners-up in the pedestrian section with this catch-phrase "Play safe and you'll live longer; look left and you'll look right." Perhaps he should have followed his own advice when out hunting wild animals.

In 1935, MacCreagh was invited to contribute to a compendium of true adventure stories entitled *Call to Adventure*, edited by Robert Benjamin.[75] The book contained fifteen short stories written by the men who had actually experienced them and it became very popular, running to several different editions. According to the publisher's promotional hyperbole " You don't read this book. You buckle your seatbelt and hang on until the last page, because packed within these pages are more adventurers, more daredevils, more rebels, more live-it-to-

---

[75] A well-known American journalist of the time who was a founding member of the Overseas Press Club of America.

the-hilt men than roam the world today." MacCreagh's contribution, entitled *Adventure and a Moral*, described some of the mishaps that he had encountered while hunting as a young man in the jungles of India and Burma. The fact that MacCreagh was asked to participate was clear evidence of his reputation and standing in the hierarchy of American adventurers and story-tellers of the time. Incidentally, his old friend, Captain Dingle was also a contributor with a tale appropriately called *Shipwrecked*.

In between his writing, MacCreagh apparently made several extended, return trips to Ethiopia during the 1930s but there are no records of the details of these journeys. According to some press reports of the time, it does seem that he met with Selassie during these visits and built up a close relationship. Indeed, according to comments made a few years later by Helen MacCreagh, "the Emperor would send a pair of elaborately harnessed mules and escort to conduct the MacCreaghs to important royal functions." She also claimed that MacCreagh introduced the very first radio to Selassie's palace, installing it personally.[76] In another press article, MacCreagh claimed to have been a financial adviser to the Emperor, though this seems very unlikely.[77] The reason for these later visits to Ethiopia, at least in part, was to prepare a new edition of his book *The Last of Free Africa*. The rising tension between Ethiopia and Italy at that time was generating a renewed interest in MacCreagh's book, both among the general public and the world's press. The author therefore released a second edition in the summer of 1935, updating the original with a new preface and epilogue to reflect the changing political situation. It's worth noting that, as a reflection of MacCreagh's standing as an author, his literary agent at the time was Nannine Joseph who also acted for several other famous people such as Franklin and Eleanor

---

[76] Article printed by *Tampa Bay News* in 1943.
[77] Boston Globe article based on an interview with MacCreagh on October 3rd, 1935,.

Roosevelt. Like the first edition, the revised version also received favourable reviews by a wide variety of publications across the USA. The September 8, 1935 *New York Times Book Review* had this to say about the new edition of *The Last of Free Africa*:

"Seven years ago Mr. MacCreagh, an incorrigible explorer into remote regions on earth, chiefly because he wants to see them, decided that Abyssinia was sufficiently remote, obscure and unknown to make it a desirable place to visit, and so he made a long stay in that country, wandered around a good deal in its interior, and wrote this book concerning what he saw and the things he learned about Ethiopia's relations with the rest of the world. Now, with Haille Selassie's realm suddenly catapulted into the very center of the world stage, and all eyes fastened upon it in painful apprehension, he has brought out a new edition, with a new preface and an epilogue in which he interprets and discusses developing affairs in the light of what he saw and learned during his long expedition."

"Take the new and old parts together, and the book makes an interesting, informed and illuminating portrayal of Ethiopia...It is a timely book that tells so many of the things the ordinary reader wants to know that no one can afford to miss it who wishes to understand the significance of what is happening... He has dug into the depths of some of the devious diplomacy that has woven webs around Ethiopia during the last few years, and the things he discloses are of even viler odor than many of the better-known achievements of European diplomatic strategy. And his comments and ways of putting what he wants to say are, to put it mildly, pungent. He, himself, calls this volume 'a ribald book.' But evidently he has written seriously and honestly what he believed to be the truth. At any rate, there is much to be learned from Mr. MacCreagh's book, both about European diplomacy and the land of Ethiopia."

MacCreagh in New York with Anazonian Wild Cat Kittens

MacCreagh Wins a New Car

A British *Adventure* Magazine Front Cover

Weird Tales **Magazine Front Cover**

A review published in the September 1935 issue of The American Foreign Service Journal was also complimentary about the second edition:

"The expedition into Abyssinia was conducted under the auspices of and made by a party of two —the author and Mrs. MacCreagh. Their adventures and their observations. with resultant conclusions, are set forth in a brilliant manner. Mr. MacCreagh has the gift of narrative and he writes with sympathy and understanding of the Abyssinians. As he remarks, there were 'a thousand absorbing things to investigate. . . . and much that one never learns of unless one digs very deeply indeed..' (MacCreagh makes) some pungent remarks, and they are all of that, on the anomalous political situation that, at present, obtains between this ancient kingdom and the nations of the world (with) a caustic discussion of the ambitions of several great powers.

Mr. MacCreagh has a high regard for 'His Imperial Highness Algaurash Tafari Makonnen,' who took at 'his coronation as Emperor the name Haille Selassie.' This high regard is expressed in the dedication of the book to Haille Selassie 'in recognition of His Great Achievements for the Progress of The Ancient Kingdom of Ethiopia.' One feels, after reading the book, that the tribute paid in the dedication is deserved."

Less than one month after these book reviews were published, MacCreagh was in Boston, on another of the many lecture tours he was now regularly undertaking in between his world travels. He had been invited to give a talk about Ethiopia to a large group of wealthy and influential people at a black-tie event in Miami. By pure chance, it turned out that he was due to speak on the very same day that Italian Fascist troops invaded Ethiopia. As a fervent advocate of Ethiopia, MacCreagh must have been both disgusted and dismayed and no doubt had to make a few last-minute changes to what he had planned to say. In the speech that he gave on that night of October 3, 1935, he dramatically forecast the certainty of Ethiopia's downfall at the hands of Italy's dictator, Mussolini. Reported in the *New York Times* the following day, the special dispatch said:

"Italy's war of conquest against Ethiopia will be over within six weeks, with the African kingdom completely subjugated, Gordon MacCreagh, adviser to and friend of Emperor Haile Selassie, tonight told an audience which was described as the wealthiest ever to assemble in a single room -- the famous Committee of One Hundred of Miami, assembled for their sixth annual Northern dinner here with a group of New England professional and industrial leaders."

The Miami Beach Committee of One Hundred was founded in 1925 to promote economic development and philanthropic service. Many famous and wealthy Americans attended its meetings, including the Vanderbilts and the Roosevelts. Despite its name, by 1935, the group had expanded to include over 400 US multi-millionaires who were joined on this particular night by an additional select group of New England industrialists. The fact that MacCreagh was invited to speak to such a highly prestigious audience, with his remarks reported on in the *New York Times*, was a significant reflection of his national reputation as an author and authority on this part of Africa. MacCreagh's prediction proved to be fairly accurate. Initially, the fighting was limited to a series of border clashes along the frontiers with Eritrea in the North and Italian Somaliland to the Southeast. However, once the full invasion got under way in the early spring of 1936, the Italians steadily pushed back the ill-armed and poorly trained Ethiopian army, taking the capital, Addis Ababa on May 5th. The Fascist army's ruthless use of tanks, artillery and poisonous gas dropped from aircraft inflicted great loss of life among both the Ethiopian military and civilians and ultimately ensured the success of Mussolini's invading forces.

In response to Ethiopian appeals, the League of Nations condemned the Italian invasion and voted to impose economic sanctions on the aggressor. Despite strong condemnation and support for the Ethiopian cause by the British, the sanctions sadly proved ineffective due to a general lack of participation by the other major powers who showed no real interest in opposing Mussolini. Emperor Haile Selassie fled into exile to Britain, along with his family and top government officials. Ethiopia was defeated, annexed and subjected to a ruthless military occupation. However, the country never officially surrendered and guerrilla resistance continued in the mountains until the Italian defeat in East Africa in 1941, during the East African Campaign of World War II.[78]

During 1935 and 1936, the League of Nations continued to ineffectually ponder and debate what to do about the Italian occupation. In June of 1936, Haile Selassie, who had featured as *Time* magazine's Man of the Year in 1935, flew from London to the League's headquarters in Geneva. He delivered an emotional, personal address in a largely unsuccessful appeal to other countries not to recognize the Italian annexation of Ethiopia. Out of fifty-two members of the League, one of the very few countries to withhold recognition of the de facto annexation was the USA.[79] I would like to think that MacCreagh's frequent lobbying on Ethiopia's behalf played a part in decision.

MacCreagh's book did not always have universally good reviews however. In the Journal of International Affairs in 1943, a reviewer called Francis Rodd [80] wrote of the revised, second edition:

---

[78] See later comments about MacCreagh's involvement in this campaign in chapter 11.
[79] Along with the USA, only five other countries withheld recognition - Mexico, China, New Zealand, Spain and the Soviet Union.
[80] Major General, Sir Francis James Rennell Rodd (1895-1978). Although the two men probably never met, they were both serving in North Africa during World War II when this review was written.

"The author and his wife, accompanied by a neither successful nor competent cinema operator . . . spent some months in Abyssinia. During this period, the first two but not the third, seem to have done some mild travelling not far from Addis Ababa. . . The descriptions of his travels in Abyssinia, of its life, of its way of travel and of its people are superficially drawn in a painfully facetious manner which palls from the outset. The one redeeming feature of the book is the last part about Christianity in Abyssinia and the policy of the Emperor. . . and his general enlightenment. . . With all the faults which the book most obviously has, and whatever he may say about the Somali and Danakil, the author has a real liking for the Abyssinians of all races."

Rodd knew the Horn of Africa well, having travelled there extensively which is presumably why he was asked to review MacCreagh's book. In 1926, he had published a book of his own about his travels in Africa and his investigation of the Tuareg tribes of the Central Sahara called *People of the Veil*. He evidently disliked MacCreagh's humorous, light-touch approach when describing his journeys and his own book is a much more serious, scholarly tome. But Rodd's critical review was a rare exception and the contribution made by MacCreagh's book to a better public understanding of Ethiopia, its people and traditions at that time was significant.

His impact on the American public's general perception of Ethiopia and its attitudes towards the country once the war between the Italians and Ethiopians began should not be underestimated. MacCreagh was one of the few Americans who had recently been in Ethiopia and knew the country well. As a result, he regularly featured in articles and interviews in newspapers across the country, from California to New Hampshire. He provided assessments on the causes of the war, comments on the conduct and progress of hostilities and predictions of the eventual outcome. However, the war in Ethiopia was not the only subject matter on which the press sought

MacCreagh's views. He was now a nationally recognized expert on African and Asian matters and occasionally found himself being consulted on other world events. For example, in an article in the January 21st, 1938 edition of the *Orlando Sentinel*, MacCreagh was asked for his views on the worsening Sino-Japanese conflict. The MacCreagh's had visited China the previous year and he clearly had strong views about the "corrupt" Chinese government and their propaganda about the war, especially in the USA.

"As Mr. MacCreagh remarked yesterday, 'China asked for trouble and got it.'. . . . It Is the Chinese students in this country and the missionaries who have come back from China who are spreading the propaganda, according to Mr. MacCreagh. He described China forsaken by a Europe who had found missionary enterprise unprofitable, and said that America, a newer country, had missionary seal which had found an answer in the country that was unoccupied by any foreign church. These missionaries are the ones now who are blaming the conflict on Japan. Mr. MacCreagh is a native Hoosier but metropolitan in his travels, for they have carried him in many strange lands. He was educated in this country and later in Aldenham and Gienalmond. Scotland, and Hanover and Gottingen. Germany. He has explored China, Borneo, South America, Australia, and other countries and conducted two explorations in Ethiopia. He rendered Halle Selassie a favor, assisted him in many wavs, and decorated by him, received an elaborate gold medal which gave him the imposing title, 'Knight of Star'. Mrs. MacCreagh said yesterday she might as well wear It for an ornament, considering that Ethiopia is a lost nation. Mr. MacCreagh, on the other hand, foretold the day when the English will fight the Italians over Ethiopia, and. having conquered It, will place Haile Selassie there as head of a protectorate. Both he and his wife spoke highly of the exiled ruler. Mr. MacCreagh has written many books, and has the pleasure of seeing his stories appear each month in current magazines. Most of these are adventure tales, and his books have included 'Big Game in the Shan States' and 'White Waters and Black'."

The article is noteworthy for the views about international politics expressed by MacCreagh as well as for the statement that he had visited Australia. There is no record or suggestion anywhere else that he ever made such a journey. This could be a mistake by the journalist who interviewed MacCreagh but more likely, it is yet another example of the author's creative mind embellishing reality, perhaps in wishful thinking. Australia was the only one of the world's continents that he didn't manage to visit. MacCreagh's comments about the deteriorating situation between China and Japan are forthright and show that he had a good awareness of world events, even if his views would ultimately place him on the wrong side of history and US government policy. His remarks about Ethiopia are both revealing and prescient. The MacCreagh's despair about the position of the country clearly runs deep. His forecast of a war between the English and Italy was totally correct and something that he would actually be a part of just a few years later.

The US press didn't just turn to MacCreagh to ask about major international events. Over the years, he was interviewed or contributed articles about all manner of topics, from snake charmers to slavery, the Chicka mud in Ethiopia, the lives of ants, hallucinogens and stimulants, the importance of wild-life and problems with the housing market. The variety of topics illustrates the breadth of MacCreagh's knowledge and interests. Again, these articles appeared in a wide cross-section of newspapers and magazines across the country as well as occasionally in overseas journals.

MacCreagh's short story output during the 1930s remained fairly constant with three or four tales published each year. This was hardly enough to provide a decent standard of living for both himself and his wife, Helen, let alone cover the cost of his periodic overseas travels. We have to assume that he made ends meet with the royalties generated by his books plus income from his lecture tours. Fortunately for MacCreagh in this respect, his popularity on the American lecture circuit remained strong. In October 1935, he was in Lafayette, Pennsylvania then in November, he addressed a packed meeting of the Athenaeum Club in Summit, New Jersey and in December he was in Massillon, Ohio. In 1936, he gave a lecture on Ethiopia to the Madison Civics Society in Wisconsin and in October that same year to Lafayette College in Pennsylvania. It's easy to see how his adventures around the world, delivered in his humorous, raconteur style, would have enthralled his American audiences, most of whom would hardly have ventured beyond their own state border. As well as writing and lecturing, MacCreagh was also meeting and corresponding with other explorers and influential people of the time on a wide variety of topics. For example, he exchanged letters on tribal customs with Vilhjalmur Stefansson, who was a well-known Canadian-born Artic explorer and Eskimo expert. Stefansson was also a member of the Adventurers' Club and became president of the New York Explorers Club in 1937. MacCreagh is recorded as meeting with Paul Bransom in 1935, a popular animal painter and illustrator from New York, presumably to provide advice about painting wild animals based on his real-life experiences.[81]

---

[81] Paul Bransom provided the drawings for the first illustrated US version of Kenneth Grahame's *Wind in the Willows*.

Chapter Eleven

Project 19

Sometime around 1937, the MacCreaghs decided to leave the New York area and permanently relocate to sunnier climes in the south. They chose St Petersburg in Florida as their new home, living initially on 16th Avenue and later moving into a newly-built modest, two-bedroom house at 2231 West Harbour Drive. It sat alongside a quiet saltwater creek and had a good-sized garden with a wide variety of trees and flowers. According to an article in the Tampa Bay Times, the MacCreaghs home was filled with furniture, souvenirs and artefacts that they had collected from their world travels.[82] It seems that the MacCreaghs quickly settled into their new life-style in the warmer Florida climate. As well as continuing with his writing and lecturing, the move allowed MacCreagh to pursue another of his hobbies – growing orchids. The plant had interested him ever since his time in the Himalayas when he had gathered sample plants for other collectors. Now he had the opportunity to cultivate his own and he had a bamboo hut built in his garden to house his growing collection. MacCreagh later became an active member of the local branch of the American Orchid Society.

He also took up sailing, joining the prestigious Big Bayou Yacht Club in St Petersburg. He evidently proved to be a capable sailor as he came second in the Club's annual handicapped race in the summer of 1939 with his boat, Ripple and was a member of the Club's race committee.[83] Perhaps he had picked up some nautical techniques from his former New York flat-mate, Captain Dingle. MacCreagh is also known to have sailed around Florida in 1940 as a guest on a US Coast Guard cutter. However, the enjoyment of their new life-style did not last

---

[82] Tampa Bay Times edition of May 24th 1942.
[83] Report in *Motor Boating* magazine, July 1939.

long. When the USA entered World War II following the bombing of Pearl Harbour in December 1941, MacCreagh soon volunteered to help the war effort, working for the US Douglas Aircraft Corporation.

In late 1941, the company had been awarded a contract by the US government to assist in the operation of a British-run aircraft assembly and repair base at Gura in Eritrea.[84] The base which was some thirty miles south of the capital, Asmara, had two primary functions. Initially, its purpose was to repair battle-damaged RAF planes involved in the North African campaign and then subsequently, it acted as an assembly and transit depot for planes shipped out in kits from the USA to be flown to Russia as part of the Anglo-American aid programme. It also later hosted an Australian field hospital. At the start, America's involvement in this project was kept secret since they were neutral and had not yet officially entered the war. Under an oath of strict secrecy, Douglas recruited volunteers from the principal U.S. airplane manufacturing centres in California, Seattle and the Midwest and at its peak, there were some two thousand American civilians employed at Gura. Each prospective employee was screened by the FBI and the highly-classified programme was given the codename of Project 19.

MacCreagh's work for Douglas and the reasons for his presence in this particular theatre of war are not clear and he never publicly commented on the details of his activities there. In a very brief comment in the Argosy magazine in January 1943, it simply recorded that 'Gordon MacCreagh is on special government duty somewhere in Africa.' Although he clearly had some knowledge of aircraft from his World War I service, he was not a skilled engineer and it's more likely that he was recruited due to his knowledge of the region.[85] We know that he spoke

---

[84] Gura had been an Italian airbase until it was captured by the advancing Allied forces.

[85] According to the *Tampa Bay Times* of April 22nd, 1962, MacCreagh acted as a foreman and interpreter.

German, Spanish and French as well as having some knowledge of local languages which would have been very useful to the Allies Some press reports indicate that after his initial stint at Gura, MacCreagh worked for both the British and American armed forces, travelling on missions to Egypt, Palestine and the Persian Gulf.

Ordinarily, MacCreagh, who was then fast approaching the age of fifty-three, would have been too old for such work. But according to an article by Harold Courlander, one of his Project 19 colleagues who came out on the same ship as MacCreagh, "he'd knocked twelve years off his age when he applied".[86] As we have already seen, this was not the first time that MacCreagh had altered his date of birth and it was to become a recurring theme on his later travels. The unavoidable ageing process clearly bothered him and he seemed to get younger with each passing year when providing his age in any official document. Usually, it's the women who lie about their age but Helen MacCreagh never seems to have joined her husband in this game.

Whatever his anticipated role was, MacCreagh was quick to volunteer and serve the Allied cause. He would also have seen this as a wonderful way to travel to a distant part of the world at someone else's expense and without the 'discipline' of actually serving in the official Armed Forces. He departed from New York in May 1942 on the *USS Chateau Thierry* transport ship, part of a convoy which sailed via Cape Town then around the Horn of Africa to Eritrea. The dangers of the war soon became apparent. According to Courlander, although MacCreagh's troop-ship arrived safely, one of the other transport vessels was sunk by a German submarine off Madagascar. Since the torpedoed ship was carrying much of the men's equipment, MacCreagh lost his precious bagpipes, much to his annoyance

---

[86] Harold Courlander's memoir *The Emperor Wore Clothes*, page 271, printed in The American Scholar magazine of Spring 1989.

As so often with MacCreagh, there is an alternative version of this event. In a 1943 interview with MacCreagh about his war-time experiences,[87] he claims that it was his ship that was torpedoed. As the vessel was sinking, "he and twelve other men clambered into a lifeboat." The German submarine then surfaced and fearing they would be machine-gunned, MacCreagh shouted out in German to the submarine crew that there were Germans in the lifeboat and so it would be unwise to shoot at them. His quick-thinking idea worked and the submarine crew let them go. After drifting through the night, they were rescued the following afternoon by an American destroyer.

It is hard to reconcile this account with that of Courlander and known facts about the *USS Chateau Thierry*. Naval records show that the ship was never torpedoed and successfully made the voyage from New York to East Africa. Since Courlander states that he came out on the same vessel as MacCreagh, one would expect him to have mentioned in his article something as important as being torpedoed. It is possible that MacCreagh changed ships during the voyage and so wasn't on the *USS Chateau Thierry* at the time of the submarine attack. It's also possible that the ship to which Courlander refers was the one that MacCreagh subsequently took after he had been torpedoed. The version of his journey provided to the reporter by MacCreagh is typically vivid as well as dramatic and certainly sufficiently detailed to seem credible.[88] However, in a similar article for the same newspaper in 1947, he said he spent three days drifting on the life-raft. Once again, I leave to the reader to decide if this was yet another of MacCreagh's real adventures or simply another marvellous, entertaining story.

---

[87] *Tampa Bay Times* article dated August 6th 1943.
[88] The essence of MacCreagh's version of events is also contained in another *Tampa Bay Times* biographical article published on September 15th, 1947.

When MacCreagh arrived in Eritrea, the Allied campaign to oust the German, Italian and Vichy-French forces from North Africa was about to begin in earnest. Following their success at the Battle of El Alamein in October 1942, the British began their steady advance westwards out of Egypt and across Libya. The rapid refurbishment of damaged RAF planes by the base at Gura played a critical role in this campaign's success. Then in November, a joint force of American and British armies landed in Morocco and began pushing east into Algeria and Tunisia. These combined operations ultimately resulted in the defeat of the Axis forces in May 1943.

According to Courlander, the two men became good friends and spent much of their off-duty hours together. MacCreagh is referred to by Courlander in another article[89] as a Scottish-American. Since Courlander was born in Indiana, one might have expected him to remark on his close friend, MacCreagh, as being a fellow Hoosier rather than a Scottish-American.[90] It is noteworthy in terms of MacCreagh's probable real origins that Courlander didn't do so. MacCreagh gained a reputation among his co-workers as something of a maverick or rebel during his time in Gura. Despite losing his cherished bagpipes on the journey out, MacCreagh managed to replace them with a set acquired from the British in Egypt. He was banned from playing them in the barracks so he went out to the edge of camp where "he piped to his heart's content." Sometimes in the evenings, he entertained the men with spellbinding tales of his various adventures in Asia and Africa. Although these stories were generally accepted as true, Courlander thought that MacCreagh "improvised and invented copiously" when narrating some of his more breath-taking events. Apparently, MacCreagh disliked discipline and his favourite pastime was "beating the system" whatever it happened to be at the time. These

---

[89] Courlander article in the US journal *Resound* in April 1987.
[90] Hoosier is the demonym for a native or resident of the State of Indiana.

articles by Courlander are interesting as they are the only first-hand accounts of MacCreagh written by someone who actually spent time with him during important events in his life. As such, they provide confirmation of many of his previously described activities and character traits.

Although Gura was a remote, off the beaten track location, the American servicemen were generally well looked after. It even had its own nine-hole golf course laid out in the desert which both MacCreagh and Courlander occasionally used. Inevitably, it had some rather unique rules:

Balls may be lifted from bomb craters and trenches without penalty.
Do not touch bombs or craters, notify authorities.
In case of air raid, the trenches are located in back of 5th and 7th greens.
Out of bounds to right of 1st, 5th and 9th holes.
If baboon steals ball, drop another ball no nearer hole— no penalty.
If ball hits an animal, play ball as it lies.

Courlander relates how at some stage in early 1943, a British airplane carrying Blatta Medhen, the Ethiopian Foreign Minister, stopped for refuelling at the Gura airport, on route to Addis Ababa. MacCreagh knew Medhen's predecessor from his previous visits to Ethiopia and the pair had enjoyed a good relationship. MacCreagh now bumped into the new Foreign Minister by pure chance at the base while his aircraft refuelled. Seizing the opportunity, MacCreagh somehow inveigled an invitation from the Minister for Courlander and himself to visit Addis Ababa to meet with Emperor Heile Selassie to discuss aid to the country now that the Italians had been expelled. Medhen was impressed to learn that MacCreagh knew the Emperor and had been awarded the Star of Ethiopia. One of the government's post-war priorities was education and as Courlander had contacts with various

American educational institutions, Medhen promised to try to arrange a meeting for them both with Selassie. This was classic, maverick MacCreagh in action. Always up for an adventure and the chance to escape the confining routine of the base. A few weeks later, with the meeting confirmed, MacCreagh and Courlander set off early one morning in an RAF plane for Addis Ababa for what was scheduled to be a simple overnight stay.

The pair duly met with Selassie and Medhen at the royal palace later the same afternoon. Courlander explored the country's educational needs while MacCreagh raised the subject of the long-proposed new dam at Lake Tana on the Blue Nile which if constructed, would make a huge difference to Ethiopia's economic prospects.[91] The audience with Selassie went well, lasting for an hour or so and he indicated that further discussions should take place the next day. Medhen then took the pair for something to eat and on to Addis Ababa airport to liaise with their pilot about arrangements for their return flight. However, on arrival, the pilot told MacCreagh and Courlander that he had been instructed to return urgently that night to Gura. The Americans could either come with him now or stay on at their own risk and have to try to sort out return transport arrangements themselves. In his typical, happy-go-lucky style, MacCreagh chose to stay and persuaded Courlander to do the same, confident that they could find their way back to Gura in a day or two.

---

[91] In his book, *The Last of Free* Africa, MacCreagh mentions the long-held desire of the British to build this dam to provide power and control the head-waters of the Nile. The concept had evidently stayed with him all these years.

Allied Airbase at Gura. Eritrea. c.1943

2231 West Harbour Drive, St Petersburg, Florida

As things turned out, their stay in Addis Ababa lasted for almost three weeks. For the first few days, the Minister for Education took them under his wing, organizing various tours and visits to existing educational facilities. Then, Minister Medhen came to their hotel to escort the Americans on a return visit to the palace to see Selassie where the subject of improving Ethiopia's education system was raised. Courlander was able to explain how he could help on his return to the USA by contacting foundations there to assist with teachers and materials for the country's schools. Selassie then turned to MacCreagh and asked him for his thoughts on the possibility of building the new dam. MacCreagh had evidently given much thought to this matter and was able to provide considerable information on what needed to be done. Courlander noted that "Selassie obviously liked Gordon" and it was clear there was a personal bond between the two men. The Emperor then asked MacCreagh "with a slight twinkle in his eyes" what he expected his remuneration to be for helping with this project. Never lost for words, MacCreagh instantly replied "Your Majesty, when the dam is under construction, I would like to have the hot dog concession."

With the official meetings now concluded, the two Americans were left to sort out their return to Gura. Since the Ethiopian government had no planes of their own at the time, they were unable to offer any help. The only aircraft flying regularly from Addis Ababa airport were operated by BOAC [92] and came under British government control. As MacCreagh and Courlander now found out, transporting American civilians around the region was very low on the priority list. They talked to BOAC officials and to the British military but to no immediate effect and with nothing else to do, the pair spent their days sightseeing and waiting. Eventually, the Minister for

---

[92] British Overseas Aircraft Corporation – the predecessor of British Airways.

Education contacted them with a solution. The government paymaster was leaving the next day by car with cash to pay the Ethiopian troops in the north of the country and could take them part way. Their transport turned out to be a vintage open-top car with the driver and paymaster in the front with bags of Maria Theresa silver dollars stuffed under their seats. MacCreagh and Courlander were squeezed into the rear seats together with the teenage son of a Baptist missionary who was on his way to school in Eritrea. They were accompanied by an open truck carrying extra fuel for the journey plus four soldiers to guard the paymaster and the money.

The small convoy set out the following morning on the mountainous climb north and it wasn't long before the veteran car began to have mechanical problems which slowed their progress considerably. MacCreagh was apparently familiar with the area and was able to guide them to a village where he remembered there was a restaurant at which they could find something to eat and overnight accommodation. Since MacCreagh's previous visit to the area, the country had been devastated by war and the restaurant was barely functioning. The meal turned out to be no more than fried eggs and their beds for the night were a few tables pushed together with an old Italian army blanket each. The ensuing four days of their journey with the paymaster were similarly difficult. They finally reached the town of Adigrat from where the two Americans and the missionary's son were able to catch a local bus to Asmara in Eritrea which passed close to the Gura base. Courlander doesn't relate what their superior's reaction was when the pair arrived back after the pair's unauthorised, extended absence.

As promised, Courlander on his return to the USA made contact with various bodies to try to develop help for improving education in Ethiopia and eventually the Guggenheim Foundation took up the cause and provided some teachers. We don't know what, if any, work was done by MacCreagh to promote the Lake Tana dam project though it seems likely he raised the topic during some of the lectures he subsequently gave on the situation in Ethiopia. However, we do know that he never obtained the hot-dog concession.[93]

As one might expect with MacCreagh, his time on 'special service' in North Africa was brought to an end in a rather dramatic manner. In the spring of 1943, he was travelling on a flight with the US Air Force when the plane he was in came under heavy enemy fire. MacCreagh was shot and wounded in the leg but the pilot managed to eventually land the damaged plane safely. According to a later newspaper report, "there were countless holes in the aircraft and one in Gordon MacCreagh." [94] Once again, it's the kind of dramatic episode that one could easily imagine featuring in an Indiana Jones film. Fortunately, the wound wasn't fatal and he subsequently spent three months recovering in an Allied hospital in Bengasi, Libya. No doubt the scar from this incident blended in well with the numerous other scars on his body that he had accumulated over the years.

Intriguingly, Courlander mentions in his memoir that MacCreagh told him he was involved in gun-running during the early years of the Ethiopian struggle against the Italian Fascist invasion and that the award of the Star of Ethiopia by the Emperor was partially in recognition of this. Since the award was made in late 1928, well before the Italians attacked in 1935, this seems on the face of it to be another example of MacCreagh's creative story-telling. However, as so often with MacCreagh's tales, it is just possible that there is an element of truth in the account – at least the gun-running part. In his book, The Last of Free Africa, he expresses much sympathy for the Ethiopian people and notes the unfairness of the arms blockade then being enforced by the Italians and French against the

---

[93] Construction of the dam was finally started in 2011 and when completed, it will be the largest hydro-electric power plant in Africa.

[94] Wellington Daily News, Kansas, January 13th 1962.

country. MacCreagh was given his prestigious award shortly after his arrival in Ethiopia with the Legendre expedition. Given his expertise with guns, knowledge of the country and contacts in the region, MacCreagh may well have been involved in arms smuggling into the country prior to, or during this visit. In this context, it's worth remembering that "Hunter Jim", the MacCreagh's guide on their first Ethiopian expedition was described in The Last of Free Africa as a gun-runner. We also know that MacCreagh made several, additional, lengthy visits to Ethiopia during the 1930s and may well have helped smuggle weapons into Ethiopia, either prior to the war with the Italians, or during it.

Chapter Twelve

Home Again

MacCreagh travelled back home from his 'special service' in late August 1943, arriving in Miami, Florida. Similar to his outbound voyage to Eritrea, his return to the USA turned out to be something of an adventure. After his recovery from his wounds in Bengasi, MacCreagh was originally due to be repatriated by troop-ship but the recent invasion of Sicily by the Allies meant that all available maritime transport had been commandeered for the campaign. The only alternative was to try to board a military flight back home but these were scarce and reserved for senior officers and urgent freight. MacCreagh's inventive (and devious) mind soon found a solution. He contacted his illustrious friend, Haile Selassie, asking him to write a letter explaining how important it was for Ethiopia and the Allied war effort for MacCreagh to return to the USA as soon as possible. The ruse worked and the Emperor's letter duly arrived together with the surprise gift of a silver dish for MacCreagh's wife, Helen. With the Emperor's supporting letter, MacCreagh was eventually able to make the return journey to Miami on a series of freight flights via Egypt, Sudan, West Africa, Brazil, British Guiana and Puerto Rico, covering a remarkable 13,000 miles in the process.

After their return to the USA, Courlander and MacCreagh lost touch with each other but a few months later, out of the blue, Courlander received a letter from MacCreagh. In it, he described how he had recently won a contract as the official photographer of the sailing races due to take place shortly in St Petersburg, Florida. MacCreagh went on to explain that he didn't have a camera and asked if Courlander could lend him his. Courlander replied that unfortunately, he had sold his camera just before leaving Gura and so couldn't help MacCreagh out. Whether or not MacCreagh went on to fulfil the photography contract is not known but it's another illustration of how carefree he was. He would jump at an opportunity, regardless of the consequences, always confident that he could somehow work his way through any problems or obstacles.

With MacCreagh's war service finished, he and Helen moved home once again, this time buying a house in the new development of Central Avenue Heights in Pinellas County of St Petersburg. Here, they settled into a new life together as local minor celebrities. MacCreagh gave occasional lectures while Helen hosted parties for local women at which she talked about her adventures and experiences while travelling overseas. It would seem that the MacCreagh's life-style was relatively simple and unpretentious. Despite having published two very successful books and written over one hundred adventure stories, MacCreagh was not a wealthy man. As we have seen, he tended to spend his money on overseas travels for the pair of them rather than personal luxuries. The formula obviously worked as the MacCreagh's relationship remained a close and happy one. MacCreagh affectionately referred to his wife as 'squirrel' which may have been a reference to Helen being more canny about money and savings than he was.

MacCreagh also resumed his writing, producing three stories in 1944 and nine in 1945. Included in the latter was a short article for Adventure magazine published in June entitled The Bloody Road to Mandalay. It is one of his rare semi-autobiographical true stories written for the pulp fiction press – MacCreagh called it a "fact story". It is a rather odd tale, both in style and content and in some ways lies closer to a church sermon than an adventure story. Although initially set during World War II at the time of the final push by the Anglo-Indian army against Japanese forces in Burma, the author soon delves into the myths and legends he learnt about when he was in the country over thirty years earlier. He takes us back to old Mandalay, the former capital of Burma and the bloody stories linked to its one-time ruler, King Mindoon. We reach the moral of the tale in the final paragraph in which MacCreagh informs us that all those who sought to possess Mandalay have shed a lot of blood and the same would happen to the Allies who were then engaged in intense fighting to rid Burma of the Japanese invaders. In the end, an unremarkable insight or prophesy and the reader can be forgiven for wondering why it was written. We do however, learn something new about MacCreagh from this "fact story". Apparently, he gained a tattoo while he was in Burma all those years ago – a splendid peacock along his whole left arm that the tattoo artist claimed was a good luck charm against snake bites. MacCreagh tells us that it obviously worked because he was never bitten by a snake.

Around this time, the MacCreaghs decided to embark on another great adventure. They sold their home in St Petersburg and left to tour the as yet, unfinished Pan-American Highway. MacCreagh apparently purchased an ancient, used hearse from a local second-hand car lot for the long journey.[95] It seems he had a predilection for old, unreliable forms of transport. The old hearse took them from Florida all the way across America and Mexico to the Guatemalan border but predictably, their chosen vehicle failed to last the entire trip. It died a death in Mexico during the return journey when an old adobe wall collapsed on top of it. Despite the unexpected end to their original plan, the MacCreaghs then changed tack, deciding to stay on in Mexico, in order to learn more about the country and its people.

They initially rented an apartment in Mexico City, but the difficulties of life in a "doll-sized apartment with limited daily water supplies" soon persuaded them to relocate to a small hacienda in the countryside. Daily life remained

---

[95] The concept of an overland road route from one end of the Americas to the other was finally agreed in 1937 by the Convention on the Pan-American Highway signed by fourteen countries and construction slowly got underway.

challenging "cooking laboriously over a charcoal fire" with only oil lamps for light and no ice but they enjoyed getting to know the locals and experiencing a different way of life. MacCreagh later pithily described Mexico as "full of delightful people, high prices and poor service." He continued to pursue his writing together with his hobby of photography but also developed a keen interest in Mayan and Aztec art and history. When the MacCreaghs eventually returned to the USA a year or so later, they brought back with them a substantial collection of Mexican artefacts to decorate their new home. The sojourn in Mexico also gave MacCreagh inspiration for several new story ideas. In 1947, he published *Xipe the Skinless*, a rather odd safari tale set in the Mexican jungle and the *Best Guide in Mexico* and *Mad Americano* both followed a couple of years later in 1950.

Back once more in St Petersburg, the MacCreaghs once again found a new home, this time near the water on Bayou Grande Boulevard. It seems that the couple's lust for overseas travel was matched by their desire to frequently live somewhere new. Then, in the summer of 1947, the couple's wanderlust returned, and they sailed to El Salvador, the smallest of the Central American countries. The MacCreagh's would have doubtless spent time enjoying the spectacular mountain scenery, especially Izalco, one of the country's volcanoes. At the time, it was erupting with such regularity that it was known as 'Lighthouse of the Pacific'. Its brilliant flares were clearly visible for great distances out to sea and at night its glowing lava turned it into a brilliant luminous cone. On the manifest of the ship that brought them home, MacCreagh was described as now being only forty-nine years old, instead of his actual age of fifty-eight, whilst Helen was listed correctly as being fifty-three. Whether this error arose from a slip of the tongue or more likely, from wishful thinking, it seems that MacCreagh was getting younger with almost every year that passed. Coincidentally, his newly revised date of birth now matched almost exactly that of the fictional Indiana Jones.

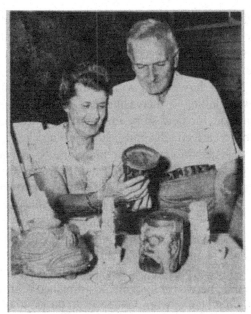

The MacCreaghs at Home in Florida

MacCreagh Working at Home (note Tattoo)

As it turned out, the trip to El Salvador was the MacCreagh's final overseas adventure. For most of the remaining seven years of his life, MacCreagh actively continued his work as a lecturer and writer, based in St Petersburg. Now however, with advancing years, MacCreagh was finding it increasingly difficult to come up with inspiration for new story ideas. After more than thirty years of creative, adventure story-telling, he had pretty well exhausted his supply of plots and scenarios. According to a local newspaper article of the time describing MacCreagh at work, he "alternatively smokes, swears, paces the floor, calls his wife for advice, plays the bagpipes or inspects one of the many relics of his wanderings." On a good day, seated at his typewriter with a clear idea, MacCreagh was able to produce around 2,000 words a day. With a typical short story usually running to some 5,000 words, he was able to create a finished draft in three days. His novelettes were longer at 20,00 to 30,000 words and so required much more time to write.

It was not only age that was catching up with MacCreagh, his readership was changing too. In the years following the end of World War II, the pulp fiction market began a marked and terminal decline, largely due to the increasing competition from comic books, television and cheap paperback novels. Also, the low-cost, pulp paper format was less suited to a more affluent post-war America which increasingly moved its reading habits upmarket. This trend would not have gone un-noticed by MacCreagh and coupled with his dearth of story-lines helps explain why he only published one or two adventure stories each year during the late 1940s. During the latter years of his life, his fictional adventure output tailed off and never recovered to his previous levels. He barely published one new story annually during the 1950s. He did however, write a very successful story called *The Emerald Crested Hoopoe* that was published by a large number of newspapers and magazines in the US and overseas during the early 1950's. Although in many ways, the tale is typical MacCreagh, mystery and adventure in what was then French Indo-China, he widely brought it right up to date by setting the story against a backdrop of the communist uprising at that time.

Although MacCreagh's writing output had significantly slowed, he continued to pursue an active schedule lecturing and commentating on Ethiopian and Asian affairs, as well as writing occasional articles for various US magazines on world affairs. One of the magazine articles was a feature for the American Gourmet Magazine[96] in 1952 on how to prepare and cook an ostrich. The magazine's up-market readers must have found this an interesting, if rather esoteric topic as very few of them would have had the chance (or even the desire) to buy and prepare such an exotic animal for dinner. MacCreagh's love of his precious bagpipes remained undiminished and in December 1951, he wrote an article for The Piper and Dancer Bulletin in which he described how to take care of a set of pipes. He also occasionally played at local concerts, parades and events and in 1952, performed in the St Petersburg Opera Society's production of the musical Brigadoon. In addition, according to his friend Bob McKnight, MacCreagh was an able guitarist with soft, haunting singing voice

In his later years, MacCreagh also turned out to be very active in another aspect of creative writing – letters to newspapers. After MacCreagh returned from his war service, he seems to have been a regular contributor to the readers' letters pages, mainly in the local Florida press. He wrote in about a wide variety of topics, from serious matters like the conduct of the war or international post-war arrangements, to more local issues such as the bad behaviour of young children and how to build cheaper houses in Florida. He clearly had strong views on these matters and his opinions were not always popular with

---

[96] First published in 1941, it was the leading USA food magazine.

other readers, as their subsequent replies to his letters revealed.

In 1953, MacCreagh wrote a story called Projection from Epsilon, which appeared in the March issue of Fantastic Adventures, the final issue of a magazine that was published in both the USA and the UK. The rather lurid cover of the magazine is an indication of how the pulp fiction press had moved down-market by the 1950s in an attempt to attract readers. Projection from Epsilon is a strange, science fiction tale about what MacCreagh calls telekdelosis or telekinesis as it is more commonly known today. Telekinesis is the supposed ability to move objects at a distance solely by mental power or other non-physical means. In the world of magic, it has links to levitation and metal-bending tricks. Although MacCreagh had previously written occasional tales of mystery or suspense, Projection from Epsilon represented his first and only venture into the world of science-fiction writing. What prompted his foray into this new genre of fiction is unclear. It may have simply been a one-off idea or perhaps he saw it as an avenue that he could develop to ward off the decline in demand for his more usual adventure stories.

After Projection from Epsilon, as far as we know, MacCreagh only wrote one more story. This was The Devil's Son which actually appeared posthumously in Adventure in August, 1955. The reason for the delayed publication in not known. However, we do know that even before Projection from Epsilon was published, MacCreagh's health had begun to deteriorate rapidly. After almost six, very difficult months in hospital, he sadly died from abdominal cancer on August 30, 1953 at the age of sixty-three. Even on his death-bed, he was still talking about his plans for new stories and his next expedition, off tramping to some new remote region. But MacCreagh's adventures were now over, at least in this world. Whatever the real facts were concerning the mysteries and enigmas of his early life, he took them all to his grave. His wife, Helen, must have been devastated to lose her close companion during so many adventures in such a traumatic manner and at such a relatively early age.

We don't know how life in old age panned out for the fictional Indiana Jones as that particular story has not yet been told – perhaps it never will be. But if such a movie is ever made, Jones' later years would probably turn out to be very similar to those of Gordon MacCreagh – a little bit of academic writing, perhaps mixed with the occasional lecture and always waiting for the excuse to disappear over the horizon on just one more exciting, go-look-see adventure.

Helen MacCreagh stayed on in Florida after Gordon's death and according to local business directories of the time, it seems she may have worked for a while in real estate.[97] Helen passed away in hospital in St Petersburg on April 22nd 1962, aged sixty-six, following a lengthy illness, probably cancer, like her husband, Gordon. As far as I have been able to establish, the couple were childless. and had no surviving close relatives, apart from Helen's younger sister, Irma Lake who lived in New Jersey. Sadly, there is no record of what happened to any of Indiana MacCreagh's papers or possessions after Helen's death. It's regrettable that none of his personal papers and correspondence files were ever archived. As for his Star of Ethiopia, it clearly meant a lot to him and I hope it was passed down and treasured within the MacCreagh's wider family somewhere.

---

[97] According to local business directories.

Chapter Thirteen

Veni, Vidi, Scripsi

Gordon 'Indiana' MacCreagh was undoubtedly a remarkable, enigmatic character who led a fascinating and incredibly adventurous life. His many achievements sit alongside his indomitable spirit, resourceful nature, enquiring mind and an enduring, irreverent sense of humour. If some of the details of his life, especially his early years, are uncertain, confusing or even conflicting, they neither diminish the man nor his outstanding achievements. Whether he was born in Scotland or the USA should not affect our admiration and respect for his many talents and abilities. MacCreagh achieved success and recognition during his life in four primary areas – as an intrepid adventurer and explorer, a big game hunter, an entertaining, easy-to-read author and as a knowledgeable commentator on contemporary world events and the lives and customs of indigenous tribespeople.

MacCreagh always admitted that he was an amateur explorer, he harboured no delusions of grandeur about his exploits or his achievements. He was neither the first nor the last of the world's fascinating characters who have sought adventure and excitement through travel. Since the very beginning of our time on earth, man has refused to be bound by the horizons that lay in front of him or the obstacles that stood in his way. MacCreagh travelled more widely than most, from the snow-capped peaks of Tibet and the Andes to the jungles of Burma or the Amazon and on to the sun-baked plains and plateaus of Africa. Like many other adventurous travellers, he experienced trials and tribulations as well as real danger. However, the indomitable manner in which he dealt with these challenging situations, usually with self-deprecating humour, marks out MacCreagh as a very special individual. He would definitely rank in my top ten list of entertaining raconteurs to invite for dinner.

As we have seen in this book, MacCreagh was an avid hunter who found pleasure in the physical and technical aspects of the chase. His many exploits in Asia and Africa trapping wild animals for zoos and circuses were highly dangerous and brought him many adventures. But he was a hunter, not a killer of wild-life. MacCreagh abhorred the practice of killing animals simply for sport, to have photos taken with their quarry or to have them stuffed for display on a wall back home. The only time he willingly killed was either to provide food 'for the pot' or in protection of his own life or that of others. He was also strongly against the "beastly" fur trade. His views were the exception then but the fact that the world has changed to largely line up with his opinions, shows he was a man ahead of his time.

Seen through the eyes of today's generation of readers, MacCreagh's tales of adventure may appear rather juvenile, lacking the slick action-drama of modern science fiction and war stories or even the Indiana Jones movies. However, facing a marauding lion alone in the bush or a tribe of head-shrinking Indians in the jungle were very real, life-threatening events that he actually experienced and successfully transposed into his numerous stories of fiction. And he did this while entertaining millions of avid readers in both the USA and overseas. As Peter Ruber commented, "One has only to read a cross-section of MacCreagh's tales to realize that he deserves high marks for storytelling. His tales have a ring of authenticity unmatched by most of his contemporaries." [98]

Gordon MacCreagh was not a man who changed the world but he certainly added to our enjoyment of it. He also improved the contemporaneous knowledge and understanding of those parts of the globe that he visited. Despite the fact that his book about the Amazon basin expedition was not written with any intention of including "a single item of scientific value", it has become part of the essential background reading for anyone interested in

---

[98] Ibid

the region. Although MacCreagh's visits to Ethiopia were primarily intended for personal travel and hunting purposes, they turned out to be very fortuitous in terms of subsequent world events. They placed him in a unique position as a knowledgeable authority on the country, allowing MacCreagh to subsequently inform and educate by means of his writing, lecture tours and press interviews.

Although MacCreagh became relatively well-known, especially during the 1920s and 1930s, he was never an 'A' list personality. He certainly mixed with the great and the famous at times, both in the USA and elsewhere but while he may have enjoyed such moments, they didn't go to his head. I don't think he was a man who sought fame and fortune or was overly ambitious. The money that he earned from his writing and lectures was only a means to an end – funding his life-long interest in travel and adventure. Neither of the two men who actually knew and wrote about MacCreagh (Harold Courlander and Bob McKnight) leaves us with the impression that he was an attention-seeker. However, I think he enjoyed his occasional moments in the spotlight – his many press interviews appear to attest to that. He also clearly got a kick out of relating his remarkable exploits to others – Courlander describes him as "spellbinding" when doing so, even if MacCreagh wasn't always totally truthful when giving his accounts.

Despite being clearly comfortable with sometimes embellishing the details of his life and adventures, one feels that this was rarely done to impress or grab attention. At heart, he was a story-teller who simply enjoyed entertaining his audience. Whilst it is easy to understand how MacCreagh could get carried away when relating his numerous, exciting adventures, it is hard to decipher why MacCreagh was so reluctant to speak about the true facts of his early years. He only ever provided brief snippets or the merest of glimpses of his childhood and family home life. There seems little doubt that his rather complicated and disjointed upbringing, probably not a very happy one, lay at the heart of this reticence. As evidenced by the early part of this book, discovering the real man and events behind the raconteur's public persona has been complicated and unusually difficult.

The complexities surrounding his early life undoubtedly helped forge his character - he was not only robust physically but mentally too. As far as I can tell, MacCreagh visited some forty different countries during his lifetime, many of them while he was still in his teens or early twenties. While his travels were undoubtedly in pursuit of his desire for adventure and to see something of the world, I believe there was a serious, deeper backdrop to his constant early roamings. There was more to it than just go-look-see. It seems to me that he was both running away or escaping from an unhappy childhood as well as searching for something. The frequent changes of jobs and locations, fascinating as they may have been, plus the actual physical challenges he chose to undergo, indicate an uncertain and unsettled mind. This constant moving-on became a series of stepping stones along the way to the underlying goal of trying to discover his true inner self.

MacCreagh was probably at his happiest when scaling high mountain peaks or tracking a tiger through dense jungle but he was by no means a loner. Those who knew him well personally, liked him and enjoyed spending time with him. He in turn, clearly enjoyed the company of others – his active participation in the many clubs or societies to which he belonged attest to that. Equally, the many photographs that he took of the various people around the world that he encountered indicate a perceptive interest in humanity, not just remote scenery and exotic animals. And of course, he was always willing to share the music from his bagpipes, even if it wasn't always appreciated.

MacCreagh was extremely fortunate to have met and married Helen, an intrepid woman who shared most of his interests and was almost as keen as her husband to go-look-see. They seem to have enjoyed an affectionate and close relationship during their marriage of almost thirty years. The couple led a comparatively modest, unostentatious life at home, particularly once they moved to Florida. Although they were minor local celebrities and happy to talk about their many fascinating experiences, this didn't go to their heads and their modest homes were always open to visitors and guests. Even when indulging their love of adventurous travelling, they certainly didn't go first class. Their limited finances simply wouldn't allow it, even if they had wanted to. For MacCreagh, an old jalopy was more fun than an ordinary car, a mule was cheaper and carried more than a horse and Shanks's pony was always more reliable than a mule.

It is clear from the various books, stories and articles written by MacCreagh that he held strong views about many of the social, environmental and political issues of the day, some of which were quite controversial at the time. Generally though, his opinions sit gently in the background of his writing and were not pushed down the throat of his readers. For example, in *The Last of Free Africa*, when the author clearly expresses his views on various contemporary social and environmental issues, he does so with a good measure of humour or pathos. More subtly, in his fictional *Kingi Bwana* tales, MacCreagh reveals his belief in the equality of man (he was strongly anti-colonial) and empathy with the local native people. Although the white man, King, is clearly the leader, he and his loyal, native sidekicks are portrayed as brother adventurers under the skin. MacCreagh's enduring interest in the life, customs and legends of the indigenous people that he met during his many travels underpin much of his literary output. Undoubtedly, his impressive foreign language skills helped him to get to know and understand such people in a way that many others couldn't.

As to the question of MacCreagh's place of birth and original nationality, the answer has to be equivocal. We shall probably never know the real truth concerning the conundrums and enigmas of MacCreagh's early life - whether he was born in Indiana or, as I believe in Scotland; his actual date of birth, whether his parents died while he was still young; where he was actually educated and why he really ended up in India. The previously published biographical accounts were largely based on MacCreagh's own input which, as we have seen, was changeable and unreliable. Also, the earliest known factual records are in fundamental and total conflict. The 1920 USA census shows MacCreagh and his parents as being born in Scotland and that he first arrived in the US in 1911. Whereas his 1921 passport application gives Indiana as his place of birth and that of his father as Chicago. Even allowing for the possibility that MacCreagh changed his name, the constantly varying versions of events throughout his life are impossible to fully understand or reconcile.

In this context, I find it odd that that there was relatively little coverage of MacCreagh's activities by the Indiana press during his lifetime. If he was truly a 'Hoosier', born in Perth, one might have expected the state's newspapers to have mentioned this and given their native son more column inches in their reports. Regardless of his actual place of birth, there is no doubt however, that MacCreagh's schooling largely took place somewhere in Britain and he was certainly of Scottish descent and heritage. At the end of the day, only MacCreagh knew the full truth.

As indicated in the very first chapter, this book is intended as not only an exploration of MacCreagh's remarkable life but also to examine along the way whether he could have been an inspiration for the fictional Indiana Jones. It is hard to be precise about the influence that Indiana MacCreagh's life and pulp-fiction writings may have had on the creators of the Indiana Jones movies, whether directly or subliminally. The strong parallels and links between the real-life MacCreagh and the cinematic character of Indiana Jones seem more than just coincidental. Although it appears unlikely that the creators of the first Indiana Jones movie knew of MacCreagh directly, both George Lucas and Philip Kaufman were keen fans of comic strips and adventure stories in their youth. They may well have read some of his stories, either in their teens or later, when they first cast around for ideas and scenarios for the film. Since MacCreagh's adventure stories largely reflect his actual life, they could well have been part of the mix that eventually produced the Indiana Jones character and some of the plot ideas.

Finally, apart from the possible connection to the fictional Indiana Jones, what is MacCreagh's overall legacy? The most obvious and tangible is his enormous literary output that provides us with a detailed insight into what he saw and did. His adventurous spirit undoubtedly lives on in his books and stories. Many of them continue to be read, reviewed and enjoyed more than half a century after his death, by both those in scientific fields or people exploring the history of adventure fiction. Less tangible perhaps but just as important, are two other things. First, is the enquiring attitude of mind he showed towards native people and their traditions that he encountered on his various overseas journeys. While many travel books have been published since his death, few have achieved a comparable, entertaining balance of fact and humour. Second, is MacCreagh's unusual attitude to travel and adventure. He clearly felt that what he saw and did was not something to be regarded as exclusive or unique. He believed anyone who really wanted to, could go on similar journeys with the minimum of fuss. No need for extensive research, suitcases crammed with supplies, expensive safaris or cheating, money-grabbing guides. For him, go-look-see was a simple, easy thing to undertake and he encouraged others to do the same. Had he lived to see the era of cheap, mass international travel that developed in the latter part of the 20th century, I'm sure he would have welcomed it.

Gordon MacCreagh is evidently still held in great respect by those who are aware of his achievements, whether in the field of exploration or through his literary work. I hope this book will go some way to raising his profile as well as bringing his remarkable life and fascinating adventures to a wider audience.

Appendix I

Gordon MacCreagh's Autobiographical Note
Published in the *Argosy* magazine in 1933.

I can't lie about my age because it's in Who's Who, and it's older than I like to believe because I aim to go world tramping as soon as I can get money enough to leave my wife at home--if she'll let me. For the old days are gone-- and wonder whether there might be eats and a camp cot wherever it might be that I would arrive.

That was the way I began. I was getting an education in a German university when I seconded a sap in a duel, and it turned out more serious than we had thought; so everybody concerned laid low for a while. Me, I had been writing to a man in Calcutta--where one shook the rupee trees and gathered wealth and glamour at the same time. This kind gentleman promised me a salary of 200 rupees per month if I would take a job in his barge business.

Well, barges were as good as anything else in the romantic East. I worked out as an under steward; and the kind gentleman gave me the job. But at the end of a month when I asked him for some rupees, he said, Oh, yes, he'd give me 200 of them per month--as soon as I head learned the barge business and was of some use to him.

So I had a fight with his son-in-law and got fired, and I took a train and went as far as it went. That was Darjeeling. I became a tea coolie driver and collected those marvelous Himalayan beetles and butterflies for a museum collector. I got into bigger stuff. Live animals for Jamrach, then the big Liverpool dealer. I understand during the war they ate them all up. I moved into the Malay islands and sent in various leopards and tigers and things. But my specialty was big snakes and orang-utans.

The war came along. I went home and lost a lot of time in a Navy training station. In the Navy I met a god called Discipline.

A couple of years sped. I sold my outstanding worth to a scientific expedition that proposed to find new and uncharted ways across South America, over ghastly Andean passes and through the whole length of the Amazon valley, which is quite a large and a wet place.

Then a spell of writing it all up. Then quite a crazy dash into Abyssinia--because nobody seemed to know anything much about it.

But my face had descended upon me by this time; and she was crazy, too, and came long--and suffered for her temerity.

Thin tent walls out in the open bush and lion noises rasper her nerves all up. And drove her crazier; so she came again the next time.

That time took us further into the interiors of things. British East Africa and Uganda borders. And we bought some mules that had been scientifically inoculated against Tsetse fly so that we wouldn't have to bother with the hideous porter safari problem. And the tsetse flies killed off half the mules anyhow, and we struggled on into bad country, and lions ate up the rest. So we had our safari after all. And the safari was tsetse speckled and ran away in heaps. And we lost baggage and were sick and the rainy season came along and caught us out in the woods; and a good time was had by all. So we came home.

Third class on a French boat to Japan--and don't you ever try that--and on to Seattle. Then we bought a used--very used--flivver and came across the continent via the auto camp routes. And we took in Columbia River Highway and Yellowstone Park and Jackson's Hole and Shoshone Canyon and Deadwood and Cody and Custer and all the places where we found all the names of our youthful reading to be honest to God true places--Dead Man's Gulch and Two Mile Bend and Snake River and Massacre Rocks and Poison Springs. And we got an awful kick out of it all.

February 18th 1933

Appendix II

Alphabetical list of Gordon MacCreagh's works

Abandoned (short story)
• *Short Stories*, Dec 1946
Abyssinian Expedition (series of articles)
• *Adventure*, Jul 1, Sep 1, Oct 1 1927, Feb 15, Apr 15, May 1, May 15, 1928
Adventure and a Moral (short story)
• *Call to Adventure*, 1935
At Long Range (short story)
• *Adventure*, Apr 1944
The Attitude of Meditation (short story)
• *Adventure*, Jan 1914
Badlands Trail (short story)
• *Adventure*, Mar 1939
Bahama Bottom (short story)
• *Adventure*, Jan 1939
Best Guide in Mexico (short story)
• *Short Stories*, Mar 1950
Big Game in the Shan States (non-fiction) c.1910
Big Jim's Way (short story)
• *Adventure*, Oct 1938
Black Drums Talking [a Kingi Bwana story] (novelette)
• *Adventure*, May 1934
Black Panther (short story)
• *Adventure*, May 1, 1935
A Blamed Amateur (short story)
• *Adventure*, Jul 1914
Blood and Steel [a Kingi Bwana story] (novelette)
• *Adventure*, Jun 1940;
• *Adventure*, Apr 1951
Blood for the Hawk (short story)
• *Adventure*, May 1952
Blood from the Blue (article)
• *Adventure*, Apr 1945
The Bloody Road to Mandalay (true story)
• *Adventure*, Jun 1945
Blow-Guns (article)
• *Adventure*, Mar 10, 1924
The Brass Idol (short story)
• *Adventure*, Oct 1913
Bugs Barton (short story)
• *Short Stories*, Feb 25, 1945

Burma Pilot (short story)
• *Short Stories*, Sep 10, 1945

The Case of the Sinister Shape [a Dr. Muncing story] (novelette)
• *Strange Tales of Mystery and Terror*, Mar 1932
• *Magazine of Horror*, Sep 1969

Consider the Camel (short story)
• *People's Favorite Magazine*, Apr 25, 1919

Contraband (short story)
• *Adventure*, Jul 1915

The Courage Medicine (short story)
• *Adventure*, Jan 10, 1925

Crackers Brat (short story)
• *Indianapolis Star*, Sep 22, 1935

The Crawling Script (short story)
• *Adventure*, Sep 30, 1923

The Creek of the Poisonous Mist (short story)
• *Adventure*, Apr 23, 1926

The Crocodile Pool (short story)
• *Adventure*, Oct 1939

The Curse of His Caste (short story)
• *All-Story Weekly*, Aug 28, 1915

Death Penalty (novelette)
• *Adventure*, Sep 1945

The Desi-Wallah (short story)
• *People's*, Feb 1916

The Devil's Son (article)
• *Adventure*, Aug 1955

The Disciple (short story)
• *Short Stories*, Aug 1914

Diver's Chances (novelette)
• *Argosy*, Dec 22, 1934

Dr. Muncing, Exorcist (novelette)
• *Strange Tales of Mystery and Terror*, Sep 1931
• *Magazine of Horror*, 1965/66

A Drum for a Warrior (novelette)
• *Adventure*, Jan 1938;
• *Adventure*, Mar 1958

Durga the Unapproachable [a "Jehannum" Smith story] (short story)

• *Argosy and Railroad Man's Magazine*, May 17, 1919
The Ebony Juju [a Kingi Bwana story] (short story)
• *Adventure*, Jul 15, 1930
The Emerald Crested Hoopoe (serialised short story)
• *syndicated to various US and foreign newspapers,* August, 1951
Everett; Commissioner of Justice (short story)
• *The Illustrated Sunday Magazine*, Jan 1915
• *Greatest Short Stories*, Collier & Sons, London, 1915
Featuring Morton St. Clair (short story)
• *Adventure*, Nov 1914
The Fish-Nets of Quoipa-Moiru (short story)
• *Adventure*, Feb 29, 1924
The Flying Chance (short story)
• *Argosy*, Dec 29, 1917
The Fool Americano (novelette)
• *Adventure*, May 1937
Forty Dollars a Foot (short story)
• *Adventure*, Sep 1915

Free to Run (short story)
• *Adventure*, Jul 1938
The Getting of Boh Na-Ghee (short story)
• *Adventure*, Nov 1913
The Ghost Tiger of Everest (short story)
• *Argosy*, Jan 30, 1937
A Good Sword and a Good Horse (short story)
• *Adventure*, mid-Jan 1921
The Great White Dayong (short story)
• *Adventure*, mid-Jun 1919
A Gun Without Guilt (short story)
• *The Saturday Evening Post,* Jan 11 1936
The Hand of Saint Ury (short story)
• *Weird Tales*, Jan 1951
He Shall Have Who Best Can Keep (novelette)
• *Adventure*, Jan 10, 1926
Heart of the Hillman (short story)
• *Argosy*, Aug 1915
The High Flier (novelette)
• *Argosy*, Mar 27, 1920

His Just Recompense (short story)
• *People's*, Dec 1916
The Honor of the Escadrille (short story)
• *Adventure*, Sep #1, 1918
Honorable Ancestor (short story)
• *Short Stories*, Dec 10, 1942;
• *Short Stories* (UK), Feb 1950
Hookey's Handicap (short story)
• *Illustrated Sunday Magazine*, May 21, 1916
I.D.B. (short story)
• *Short Stories*, Feb 10, 1933:
• *Short Stories* (UK) early-Sep 1933
The Inca's Ransom (novelette)
• *Adventure*, Jul 10, 1924
In the Lap of Enigma (short story)
• *Illustrated Sunday Magazine*, July, 1913
The Ivory Killers [a Kingi Bwana story] (short story)
• *Adventure*, Jun 1933
The Jade Hunters (short story)
• *Adventure*, 1912

"Jehannum" Smith [a "Jehannum" Smith story] (short story)
• *Argosy*, and Railroad Man's Magazine Apr 5, 1919
The Jest of the Jungle (short story)
• *Adventure*, Aug 20, 1925
Job Across Jordan (short story)
• *Adventure*, Jul 1946
Jungle Business (short story)
• *Adventure*, Jun 10, 1925
Jungle Feud (short story)
• *Maclean's,* Canada, March 15, 1941
Jungle Hate (short story)
• *Short Stories*, Jan 1950
The Jungle Never Tells (short story)
• *Adventure*, Oct 1, 1932
Jungle Raider (short story)
• *Short Stories*, Nov 10, 1944
The King of the Black Water (short story)
• *Adventure*, Aug 8, 1926
King's Sons (short story)
• *Adventure*, Mar 1, 1933

Kite Dickinson—Non-Combatant (short story)
• *Adventure*, Aug 1915
The Last of Free Africa (non-fiction)
• Century Company, New York, Sep 1928.
The Last of the Monopoles (short story)
• *People's*, Apr 1917
The Leopard Trap (short story)
• *People's*, Sep 1916
Lion Trouble (non-fiction)
• *The Blue Book Magazine*, Jul 1936
Lion Trouble (novelette)
• *Adventure*, Jan 1945;
• *Adventure* (Canada), Jun 1945
The Lions That Stopped a Railroad (article)
• *Adventure*, Mar 1938
Little Fox Trails (short story)
• *Short Stories*, with Keith's House Plans Jun 1915
Little Red Devils (short story)
• *Short Stories*, Jun 25, 1940;
• *Short Stories* (UK), Oct 1941

The Lord of Elephants [a Neil MacNeil story] (short story)
• *People's*, Aug 1916
The Lost End of Nowhere [a Kingi Bwana story] (novelette)
• *Adventure*, Jan 15, 1931
MacNeil's Job [a Neil MacNeil story] (short story)
• *People's*, Mar 1916
Mad Americano (short story)
• *Short Stories*, May 1950;
• *Short Stories* (UK) May 1953
The Mad Hakim (short story)
• *Adventure*, Sep 1933
Mamu the Soothsayer (novelette)
• *Adventure*, May 20, 1924
The Man from Home (short story)
• *Adventure*, May 1915
A Man to Kill [a Kingi Bwana story] (short story)
• *Adventure*, Nov 1938

The Man Who Didn't Care (short story)
• *The Popular Magazine*, Feb 20, 1919
• *The Popular Magazine*, Dec 1, 1929

A Man's Price (short story)
• *Short Stories*, Jun 25, 1942;
• *Short Stories* (UK) Oct 1948

Masterless Man (novelette)
• *Short Stories*, Nov 1949

A Matter of Class (short story)
• *Argosy*, Feb 16, 1918

Matto Grosso Fury (short story)
• *Jungle Stories*, Vol. 4, No. 11, 1950

McGrath's Job (short story)
• *Argosy*, Aug 23, 1919

The Miracle of Kali (short story)
• *Illustrated Sunday Magazine*, Apr 18, 1915

Missing in M'Bwemo (short story)
• *Short Stories*, May 10, 1945

The Monkey God (short story)
• *All Around Magazine*, Jul 1916

Mountain Promise (short story)
• *Adventure*, Aug 1942

The Much Maligned Army Ant (article)
• *Adventure*, Mar 30, 1924

The Mud Fort o'Tongsa (short story)
• *People's*, Mar 1917

Naked Men of Naga [a "Jehannum" Smith story] (novelette)
• *Argosy*, Apr 17, Apr 24, May 1, 1920

Nobody Kills Don Guzman (short story)
• *Adventure*, Nov 1937;
• *Adventure*, Feb 1962

The Oath by the Earth (short story)
• *Adventure*, Oct 1915

An Officer and a Gentleman (short story)
• *Adventure*, Oct #1, 1917

The Old Oil (novelette)
• *Adventure*, Feb 1946

One Bottle with Ears (novelette)
• *Short Stories*, Jan 25, 1940;
• *Short Stories* (UK) Jul 1940

Out of the Jungle (short story)
• *Adventure*, Mar 30, 1925

Pacifist (short story)
- *Street & Smith's Complete Stories*, Dec 15, 1933

Pearly Depths (novelette)
- *Argosy*, Nov 17, 1934

Picaroon (short story)
- *Short Stories*, Aug 25, 1944;
- *Short Stories*, Incorporating West (UK) Oct 1954

The Pied Piper of Nairobi (short story)
- *Adventure*, Nov 1941

The Poison Devil (short story)
- *Adventure*, Feb #1, 1919

Potent Paint (short story)
- *Adventure*, Nov 1945

Projection from Epsilon (short story)
- *Fantastic Adventures*, March 1953

Quill Gold [a Kingi Bwana story] (short story)
- *Adventure*, Feb 1, 1931

Raiders of Abyssinia [a Kingi Bwana story] (short story)
- *All Aces*, Apr 1936

The Rajah's Royalty [a Neil MacNeil story] (short story)
- *People's*, Jul 1916

The Rat of Cayenne (short story)
- *The Danger Trail*, May 1928

Reptile Man (sl)
- *Argosy*, Feb 11, Feb 18, 1933;
- *Argosy* (Canada) Feb 25, 1933

Reward (short story)
- *Adventure*, Jun 15, 1927

The Rope of Pedro Mendez (short story)
- *Adventure*, Jan 10, 1924

The Safari of Danger (novelette)
- *Argosy*, Feb 10, 1934

Safe and Sane (short story)
- *Adventure*, Jun #1, 1918

Sanford Hale, Aviator (short story)
- *Munsey's Magazine*, Oct 1918

The Scarecrow (short story)
- *Adventure*, Nov 1915

Sense of Balance (short story)
- *Adventure*, Aug #1, 1918

The Slave Runner [a Kingi Bwana story] (novelette)
• *Adventure*, Apr 1, 1930
Slaves for Ethiopia [a Kingi Bwana story] (short story)
• *Adventure*, Jul 1939;
• *Adventure*, Jan 1951
The Society of Condors (short story)
• *Adventure*, Apr 1, 1927
A Soldier Salutes the Uniform (novelette)
• *Adventure*, Jun 1941
The Soul Winder [a Neil MacNeil story] (short story)
• *People's*, Jun 1916
The Spoils of War (short story)
• *People's*, Oct 1916
The Spring Running [a Neil MacNeil story] (short story)
• *People's*, May 1916
A Stone in a Sling (short story)
• *Collier's Weekly*, Jan 17, 1942
Strangers of the Amulet [a Kingi Bwana story] (serial)
• *Adventure*, Apr 15, May 1, 1933

Strong as Gorillas [a Kingi Bwana story] (short story)
• *Adventure*, Apr 1940
Tact and Some Diplomacy [a Jehannum Smith story] (novelette)
• *Argosy*, Dec 13, 1919
Tarred with the Trader Brush (short story)
• *Adventure*, Feb 1941
The Tenderfoot of Nairobi (short story)
• *Adventure*, Dec 1940
"That Blasted Discipline" (short story)
• *Adventure*, Sep #1, 1917
Tiger's Orphan (short story)
• *Adventure*, Dec 15, 1931
The Trail Smellers (short story)
• *Adventure*, Dec 20, 1924
Unprofitable Ivory [a Kingi Bwana story] (novelette)
• *Adventure*, Jun 15, 1931
The Unwilling Passenger (short story)
• *People's*, Nov 1916

Vegetable Ivory (syndicated short story)
• *Sunday Free Press*, Jan 1913

Wardens of the Big Game [a Kingi Bwana story] (novelette)
• *Adventure*, Apr 15, 1935

The Weeping Buddha (short story)
• *People's*, Feb 1917

When a Man Must (short story)
• *Short Stories*, May 10, 1948

The White Man of Borneo (short story)
• *Adventure*, Oct #1, 1918

White Waters and Black (non-fiction)
• Grosset & Dunlap, New York, 1926

Willful Non-Cooperation (novelette)
• *Short Stories*, Jul 10, 1945

The Witch Casting [a Kingi Bwana story] (short story)
• *Adventure*, Nov 1, 1931

The Wood Devil Thing (short story)
• *People's*, Jan 1916

Worshipers of Boondi [a "Jehannum" Smith story] (short story)
• *Argosy*, Nov 8, 1919

Xipe the Skinless (novelette)
• *Adventure*, May 1947

Y' Gotta Show Me! (short story)
• *All-Story Weekly*, Apr 8, 1916

Zimwi Crater (novelette)
• *Argosy*, Aug 11, 1934

Appendix III

Gordon MacCreagh's works by first date of publication

Big Game in the Shan States c.1910
The Jade Hunters 1912
Vegetable Ivory Jan 1913
In the Lap of Enigma July, 1913
The Brass Idol Oct 1913
The Getting of Boh Na-Ghee Nov 1913
The Attitude of Meditation Jan 1914
A Blamed Amateur July 1914
The Disciple Aug 1914
Featuring Morton St. Clair Nov 1914
Everett; Commissioner of Justice Jan 1915
The Miracle of Kali Apr 18, 1915
The Man from Home May 1915
Little Fox Trails Jun 1915
Contraband July 1915
Heart of the Hillman Aug 1915
Kite Dickinson—Non-Combatant Aug 1915
The Curse of His Caste Aug 28, 1915
Forty Dollars a Foot Sep 1915
The Oath by the Earth Oct 1915
The Scarecrow Nov 1915
The Wood Devil Thing Jan 1916
The Desi-Wallah Feb 1916
MacNeil's Job Mar 1916
Y' Gotta Show Me! Apr 8, 1916
The Spring Running May 1916
Hookey's Handicap May 21, 1916
The Soul Winder Jun 1916
The Monkey God July 1916
The Rajah's Royalty July 1916
The Lord of Elephants Aug 1916
The Leopard Trap Sep 1916
The Spoils of War Oct 1916
The Unwilling Passenger Nov 1916

His Just Recompense Dec 1916
The Weeping Buddha Feb 1917
The Mud Fort o'Tongsa Mar 1917
The Last of the Monopoles Apr 1917
An Officer and a Gentleman Oct 1917
That Blasted Discipline Sep 1917
The Flying Chance Dec 29, 1917
A Matter of Class Feb 16, 1918
Safe and Sane Jun, 1918
Sense of Balance Aug 1918
The Honor of the Escadrille Sep1, 1918
Sanford Hale, Aviator Oct 1918
The White Man of Borneo Oct 1918
The Poison Devil Feb 1919
The Man Who Didn't Care Feb 20, 1919
Jehannum Smith Apr 5, 1919
Consider the Camel Apr 25, 1919
Durga the Unapproachable May 17, 1919
The Great White Dayong Jun 1919
McGrath's Job Aug 23, 1919
Worshipers of Boondi Nov 8, 1919
Tact and Some Diplomacy Dec 13, 1919
The High Flier Mar 27, 1920
Naked Men of Naga Apr 17, 1920
A Good Sword and a Good Horse Jan 1921
The Crawling Script Sep 30, 1923
The Rope of Pedro Mendez Jan 10, 1924
The Fish-Nets of Quoipa-Moiru Feb 29, 1924
Blow-Guns Mar 10, 1924
The Much Maligned Army Ant Mar 30, 1924
Mamu the Soothsayer May 20, 1924
The Inca's Ransom Jul 10, 1924
The Trail Smellers Dec 20, 1924
The Courage Medicine Jan 10, 1925
Out of the Jungle Mar 30, 1925
Jungle Business Jun 10, 1925

The Jest of the Jungle Aug 20, 1925
He Shall Have Who Best Can Keep Jan 10, 1926
The Creek of the Poisonous Mist Apr 23, 1926
The King of the Black Water Aug 8, 1926
White Waters and Black 1926
The Society of Condors Apr 1, 1927
Reward Jun 15, 1927
Abyssinian Expedition 1927-28
The Rat of Cayenne May 1928
The Last of Free Africa Sep 1928
The Slave Runner Apr 1, 1930
The Ebony Juju Jul 15, 1930
The Lost End of Nowhere Jan 15, 1931
Quill Gold Feb 1, 1931
Unprofitable Ivory Jun 15, 1931
Dr. Muncing, Exorcist Sep 1931
The Witch Casting Nov 1, 1931
Tiger's Orphan Dec 15, 1931
The Case of the Sinister Shape Mar 1932
The Jungle Never Tells Oct 1, 1932
King's Sons Mar 1, 1933
I.D.B. Feb 10, 1933
Reptile Man Feb 11, 1933
Strangers of the Amulet Apr 15, 1933
The Ivory Killers Jun 1933
The Mad Hakim Sep 1933
Pacifist Dec 15, 1933
The Safari of Danger Feb 10, 1934
Black Drums Talking May 1934
Zimwi Crater Aug 11, 1934
Pearly Depths Nov 17, 1934
Diver's Chances Dec 22, 1934
Adventure and a Moral 1935
Wardens of the Big Game Apr 15, 1935
Black Panther May 1, 1935
Crackers Brat Sep 22, 1935
A Gun Without Guilt Jan 11 1936
Raiders of Abyssinia Apr 1936
Lion Trouble July 1936
The Ghost Tiger of Everest Jan 30, 1937

The Fool Americano May 1937
Nobody Kills Don Guzman Nov 1937
A Drum for a Warrior Jan 1938
The Lions That Stopped a Railroad March 1938
Free to Run July 1938
Big Jim's Way Oct 1938
A Man to Kill Nov 1938
Bahama Bottom Jan 1939
Badlands Trail Mar 1939
Slaves for Ethiopia Jul 1939
The Crocodile Pool Oct 1939
One Bottle with Ears Jan 25, 1940
Strong as Gorillas Apr 1940
Blood and Steel Jun 1940
Little Red Devils Jun 25, 1940
The Tenderfoot of Nairobi Dec 1940
Tarred with the Trader Brush Feb 1941
Jungle Feud March 15, 1941
A Soldier Salutes the Uniform Jun 1941
The Pied Piper of Nairobi Nov 1941
A Stone in a Sling Jan 17, 1942
A Man's Price Jun 25, 1942
Mountain Promise Aug 1942
Honorable Ancestor Dec 10, 1942
At Long Range Apr 1944
Picaroon Aug 25, 1944
Jungle Raider Nov 10, 1944
Lion Trouble Jan 1945
Bugs Barton Feb 25, 1945
Blood from the Blue Apr 1945
Missing in M'Bwemo May 10, 1945
The Bloody Road to Mandalay Jun 1945
Willful Non-Cooperation Jul 10, 1945
Death Penalty Sep 1945
Burma Pilot Sep 10, 1945
Potent Paint Nov 1945
The Old Oil Feb 1946
Job Across Jordan July 1946
Abandoned Dec 1946
Xipe the Skinless May 1947
When a Man Must May 10, 1948

Masterless Man Nov 1949
Jungle Hate Jan 1950
Matto Grosso Fury 1950
Best Guide in Mexico Mar 1950
Mad Americano May 1950
The Hand of Saint Ury Jan 1951
The Emerald Crested Hoopoe August 6, 1951
Blood for the Hawk May 1952
Projection from Epsilon March 1953
The Devil's Son Aug 1955